Exploring Careers in Construction

National Center for
Construction Education and Research

Prentice Hall
Upper Saddle River, New Jersey   Columbus, Ohio

Prentice-Hall, Inc.
Upper Saddle River, New Jersey 07458

This information is general in nature and intended for training purposes only. Actual performance of activities described in this manual requires compliance with all applicable operations procedures under the direction of qualified personnel. References in this manual to patented or proprietary devices do not constitute a recommendation for their use.

Printed in the United States of America

10 9 8 7 6

ISBN: 0-13-910159-4

Prentice-Hall International (UK) Limited, *London*
Prentice-Hall of Australia Pty. Limited, *Sydney*
Prentice-Hall of Canada, Inc., *Toronto*
Prentice-Hall Hispanoamericana, S. A., *Mexico*
Prentice-Hall of India Private Limited, *New Delhi*
Prentice-Hall of Japan, Inc., *Tokyo*
Simon & Schuster Asia Pte. Ltd., *Singapore*
Editora Prentice-Hall do Brasil, Ltda., *Rio de Janeiro*

Preface

Every American lives in the built environment—the product of the construction industry. This book is intended to introduce you to the world of construction and the people who build America.

On the average, the construction industry employes more than five million people. In the coming years, hundreds of billions of dollars will be spent on the construction and maintenance of industrial plants, roads, and bridges. More than one trillion dollars will be spent on construction materials, labor, and services. Hundreds of thousands of new jobs will be opening in construction, from carpenter to supervisor to architect.

There is a world of opportunity in the construction industry. You are invited to explore the many important roles you could play in construction and to consider a career in this dynamic industry.

For more information about careers in the construction industry, contact the National Center for Construction Education and Research at (352) 334-0911.

Acknowledgments

Thanks are extended to the following champions of the industry for their contributions to this book:

Shirley Busbice

Steve Greene

Ellie Hein

Roosevelt Tillman

Table of Contents

CHAPTER 1 — ***INTRODUCTION*** 5
- Growth of the Construction Industry 5
- Construction Education Opportunities 5
- Apprenticeship 7
- Constructing a Building 10
- Contractors 21
- Employment Opportunities 22

CHAPTER 2 — ***ORGANIZATION & STRUCTURE*** 27
- General Contractor 30
- Private and Public Construction 31
- Private and Public Construction 31
- Company Policies and Procedures 31
- Entering the Contracting Business 33

CHAPTER 3 — ***WORK HABITS*** 37
- Starting a New Job 37
- Employee Expectations 38

CHAPTER 4 — ***CONSTRUCTION SAFETY*** 45
- Signs 45
- Scaffolds 60
- Electric Shock 60
- Safety and Your Income 62

CHAPTER 5 — ***THE BUILDING TRADES*** 63
- Carpenters 63
- Drywall Installers 64
- Electricians 66
- Glaziers 69
- HVAC Technicians 70
- Industrial Coaters 75
- Insulation Workers 75
- Instrumentation Workers 76
- Ironworkers 76
- Masons 78
- Metal Building Assemblers 79
- Millwrights 80
- Painters 80
- Pipefitters 81
- Plumbers 82
- Sheet Metal Workers 84
- Welders 85

CHAPTER 6 — ***SPECIALIZED TRADES*** 87
- Floor Finishers 87
- Roofers 88
- Demolition Workers 89
- Elevator Constructors 90

Chapter 1
Introduction

GROWTH OF THE CONSTRUCTION INDUSTRY

During the past 50 years, the construction industry has been a leading provider of job opportunities in the nation. During the 1940s, new construction spending totaled about $8 billion annually. By 1987, spending had grown to over $400 billion. The construction industry has provided goods and services that account for about 15 percent of the gross national product during the last half century; and at the close of World War II, this industry was providing more goods and services than any other industry in the United States.

As with any industry, times and technology change. New tools and equipment have been developed that help construction workers build structures faster, saving the public millions of dollars in labor. Foundations that used to be dug by hand are now dug by heavy earth-moving equipment. As a result, residential, commercial, and industrial construction have prospered.

The materials used to build many homes and establishments less than 20 years ago are now obsolete, and computers have taken the industry into another era. Plans can now be made faster, and the right types of materials can be ordered and on the job when they are needed. Constructing a building today may be compared to completing a jigsaw puzzle. The border of the puzzle is like the footings and foundation of a building. Once they are in place, the various trades work together to fill in the missing pieces until the puzzle is complete.

CONSTRUCTION EDUCATION OPPORTUNITIES

Millions of dollars have been spent to research and design new equipment and to establish education programs, for the industry has recognized that a strong education foundation for its workers is as important to the industry's long-term success as the strength of a building foundation.

The future of the industry depends largely on the establishment of an effective, nationwide education and training program in the various building trades. A variety of programs exist today and new ones are being developed to meet the challenges of technological advances in the industry as they occur.

Education Programs

The educational process for the building trades can begin in high school. High school vocational education allows students to learn about a variety of careers, including the construction trades, by placing them in jobs for part of the school day.

Many high school students are uncertain about what they want out of life, but they are inclined to

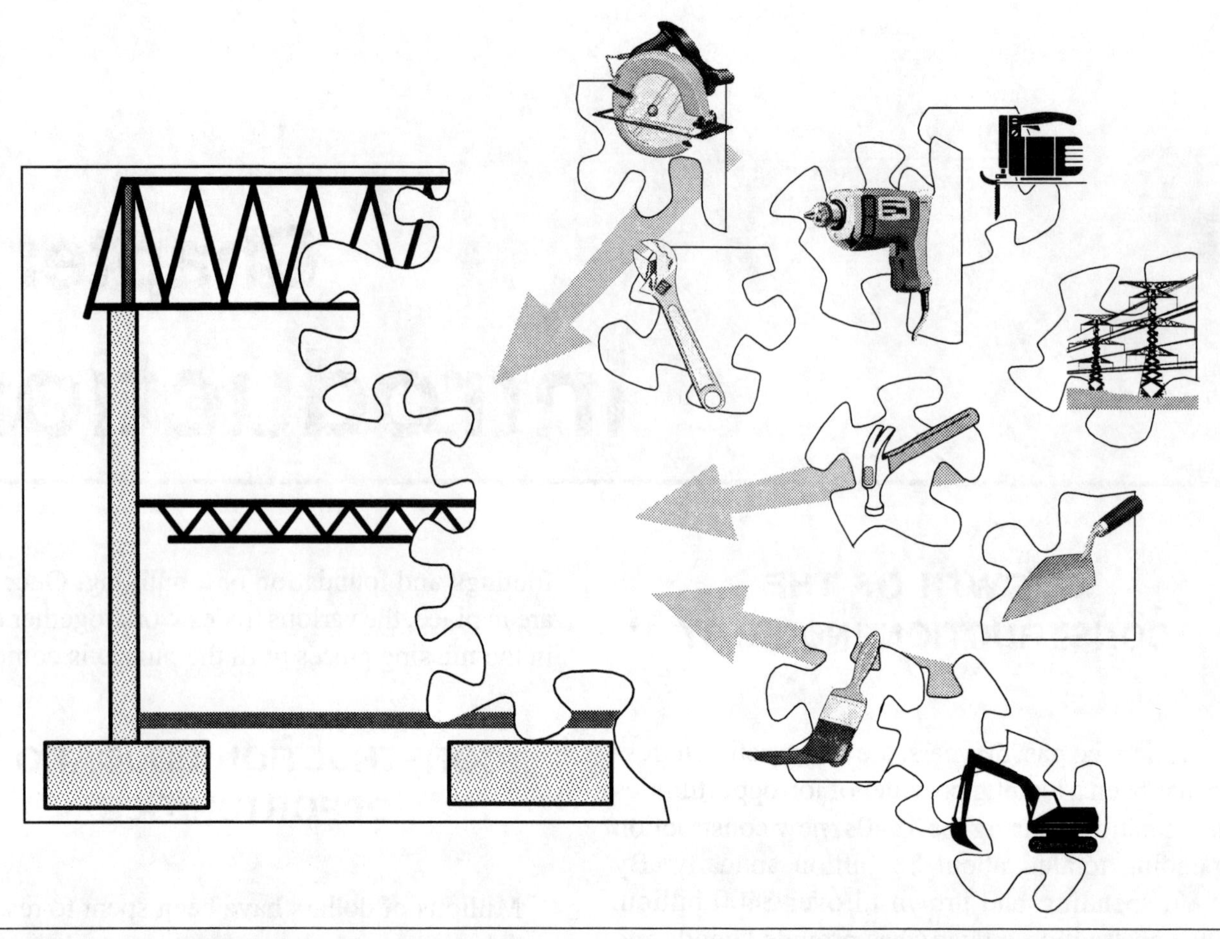

The construction of a building today may be compared to completing a jigsaw puzzle.

accept change and new ideas. Students who study construction will be surprised by this fascinating and intricate business and the vast array of trades available to them. Some people find that they are natural athletes; others take to electrical wiring like ducks to water. The point is to explore. There may be a trade out there that you will love.

Vocational construction programs at the college level offer more intense and trade-specific training, and many colleges offer both undergraduate and graduate degree programs. Although programs vary from state to state, each provides the necessary tools for becoming a skilled construction worker. The University of Colorado, for example, offers an undergraduate degree in the building trades through its College of Engineering and Applied Sciences. Arizona State University offers a program through its construction department.

Four-year construction management programs include such courses as accounting, construction methods, engineering, estimating and cost control, mathematics, materials, planning and scheduling, construction law, and human relations.

More than 200 two-year technical schools or junior colleges offer programs for construction technicians—estimators, drafters, supervisors, or field engineering assistants who can work with architects, engineers, contractors, and operating employees.

Many organizations and associations also offer courses for prospective construction workers in the building trades.

In the past, construction workers worked their way up through the trades to become experts, but many trades now require some formal education or training from an accredited institute or apprenticeship program (see figure 1-1). Those already working in a trade find that they need continuing education on the new products and materials constantly being developed.

Many experienced trade workers look to workers coming out of training programs for education about new technology in the industry. In exchange, experienced workers have a vast amount of practical experience to share with the inexperienced, educated construction worker. The experienced, educated building and construction worker today will rely heavily on future generations.

The construction industry needs a trained and educated work force to meet the ever-increasing demands of future construction projects. This work force will require classroom training as well as on-site experience. Everyone wants financial security; the time to begin your training for a rewarding career is now. As in other jobs, you may start at the bottom of the ladder in construction, but with quick wits, simple observation, and listening, you will find the construction industry a very rewarding place to be.

Training programs are direct routes to a dynamic career. They are not easy by any means, but the financial rewards and deep personal satisfaction they offer are worth the effort.

APPRENTICESHIP

Classroom training and on-site experience go hand in hand. The beginner—the apprentice—needs both to have a rewarding career.

History of Apprenticeship

Apprenticeship training transfers practical knowledge from one generation to another, and this method of learning a trade goes far back in human history. Provisions for teaching apprentices are found in the Babylonian Code of Hammurabi, which is 4000 years old. In ancient Rome, the high level of technical achievement that produced aqueducts, great public buildings like the Coliseum, and a large network of roads was due to craftspeople and artisans whose skills were handed down from father to son.

One ancient contract provided that a man named Heraclas be taught weaving during a period of five years. He was given food, 20 annual holidays, and a new tunic each year. After two and a half years, he was paid 12 drachmas; in his fifth year he made twice as much.

During the Middle Ages (500 A.D. to about 1500 A.D.), two social classes developed: merchants and skilled artisans. Both organized themselves into guilds in order to reduce competition and provide training for apprentices. In these training programs, apprentices were indentured (bound by contract) to a master craftsperson who agreed to train them in the craft and provide them with food, clothing, and shelter. Apprenticeships lasted from 2 to 10 years, depending on the trade.

Modern Apprenticeship Training

Apprenticeships remained virtually unchanged until the education reforms of the 19th century, when industrial education became a formal part of the public school system. Compensation changed from food, clothing, and shelter to payment of wages.

Apprenticeship systems gradually developed for a host of new industries that rely on machinery and electrical equipment. The graduated wage scale first appeared in the mid-1800s. In 1911, Wisconsin enacted the first legislation in the United States to regulate apprenticeship systems. This law placed apprenticeships under the authority of an industrial commission.

In the 1920s, national employer and labor organizations, educators, and government officials began an effort to initiate a national apprenticeship system. The construction industry was a prime force in this movement. The need for comprehensive training increased after World War I, when immigration restrictions slowed the flow of skilled foreign workers.

In 1934, Congress passed the Fitzgerald Act—the first step in forming a national system of ap-

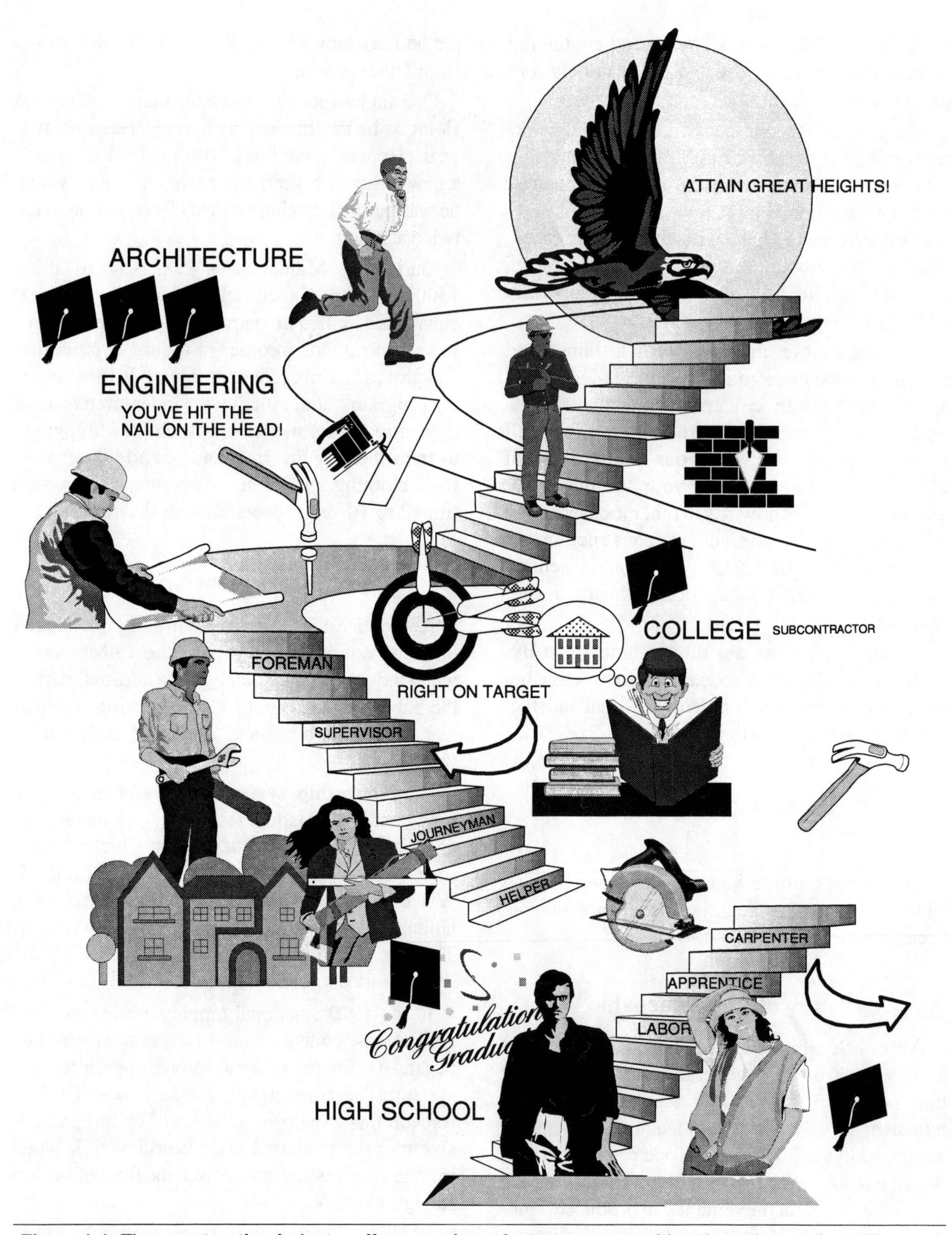

Figure 1-1: The construction industry offers a variety of career opportunities that account for millions of jobs in dozens of different fields.

Congratulations Graduate
VOCATIONAL SCHOOL

prenticeship training. The act established the Bureau of Apprenticeship and Training (BAT), which has since worked closely with employers and labor groups, inspection departments, colleges, vocational schools, and state agencies to promote apprenticeship programs and advise these groups about such training.

Programs registered with BAT must meet the following provisions:

- Apprenticeship opportunities must be available to all.
- Training must be combined with on-the-job experience.
- A minimum of 144 hours per year of related training must be provided.
- The apprenticeship must offer an increasing schedule of wages.
- Apprentices must have proper supervision in adequate facilities.
- Job performance and related instruction must be periodically evaluated.
- Apprentices who successfully complete the program must be formally recognized.

The apprenticeship system is still changing. Today, apprenticeships respond to technological and social conditions brought about by scientific discoveries, new teaching methods, expanding industry, and a booming population. As a result, apprenticeships have been set up in new trades and updated in many older ones.

Advantages of Apprenticeship

Apprenticeship is an efficient way for workers to learn skills because the training is planned and organized. Also, apprentices earn as they learn.

Industry benefits as well. From apprenticeship programs come craft workers who are competent in all aspects of their trade and are able to work independently. When changes are made in production, these workers are able to change work components quickly to suit the changing needs. Quality workers with these skills are vital to industrial progress.

Apprenticeship training provides a ready source of skilled workers to meet current and future employment needs. The on-the-job training component provides the practical skills for the trade. The classroom instruction provides the background information required to understand the trade—what it involves and how the job gets done.

Trade workers who advance in knowledge and skill stay productive and contribute to the economic success of the firm they work for. Individual initiative is doubly important: workers move up the ranks to positions of greater responsibility and the company remains competitive.

CONSTRUCTING A BUILDING

Constructing a building takes many hours of planning, financing, design work, and understanding plans and specifications before the first shovel hits the dirt. The process requires a closely coordinated team of experts.

Architects

In most cases, the first step in constructing a building is to hire an architect to prepare the drawings and specifications for the project.

The owner of the building meets with the architectural firm and describes the type of building wanted, how the building will be used, the budget for the project, and other important information. During this preliminary conference, the architect makes recommendations to the owner and decides if the project is feasible.

Once the initial information is obtained, the architect and staff (including structural, electrical, and mechanical consulting engineers) research the project thoroughly and sketch preliminary drawings for the owner to review. These drawings will eventually become the working drawings and written specifications for the project. The drawings usually include the following:

- A plot plan showing the location of the building and utilities on the property (see figure 1-2).
- Elevations of all outside faces of the building.
- Floor plans showing the walls and partitions for each floor or level.
- Vertical cross-section views of the building to show the various floor levels and details of the foundation, walls, floors, ceilings, and roof construction.
- Large-scale drawings showing any other details of construction that may be required.
- Equipment and material schedules.

Once these drawings are approved, the architect often gets bids for the building owner from general contractors and helps the owner decide which contractor to choose. The architect might also represent the owner during construction, inspecting the work to make sure that it is being done according to the requirements of the architect's plans.

Depending on the size and complexity of the project, approval of the preliminary drawings may take as little as one day to as long as several months or years.

Survey and Plot Plan

Once land has been purchased, a civil engineer or land-surveying firm is usually hired to survey the land on which the building will be built. During this process, the surveyor marks the legal property boundaries and gives the architect or owner a certified plot plan (drawing of the building site) that shows some or all of the following details:

- Property boundaries.
- Existing streets, sidewalks, buildings, and trees.
- Existing utilities (telephone, electric, cable television, and water, and drainage lines).
- Shape of the land.

Once all drawings have been completed and approved by the owner, the architect or planning engineer designs the building. The shape and size of the building is drawn to scale and placed on the surveyor's plot plan at the desired location. The architect's plan also often shows how the shape of the land must change in order to accommodate the new building.

Floor Plans

Floor plans are usually the first architectural drawings to be prepared. These plans show the outline and all details as they appear when looking directly down on the building. They show two dimensions—length and width. The floor plan of a building is drawn as if a slice was taken through the building—about window height—with the top portion removed to reveal the part underneath (see figure 1-3). If the building has more than one floor, other slices may be taken at varying distances from the ground to show the other floors.

Let's say that we wanted a plan view of a commercial laundry. The plan shows the building cut away from about the middle of the first-floor windows. By looking down on the uncovered portion, we can see every detail, partition, and window and door opening. This drawing is called the *first-floor plan.* A cut through the second-floor windows (if applicable) would be the *second-floor plan*, and so on. A single-floor building, as shown in figure 1-3, will only have one basic floor plan (figure 1-4), while a high-rise office building may contain a dozen or more floor plans. Each floor plan is usually copied for each separate trade; in other words, there is an electrical floor plan; a plumbing floor plan; a heating, ventilating, and air-conditioning (HVAC) floor plan; and so on.

Note that the floor plan in figure 1-4 not only shows the location of all partitions, windows, and

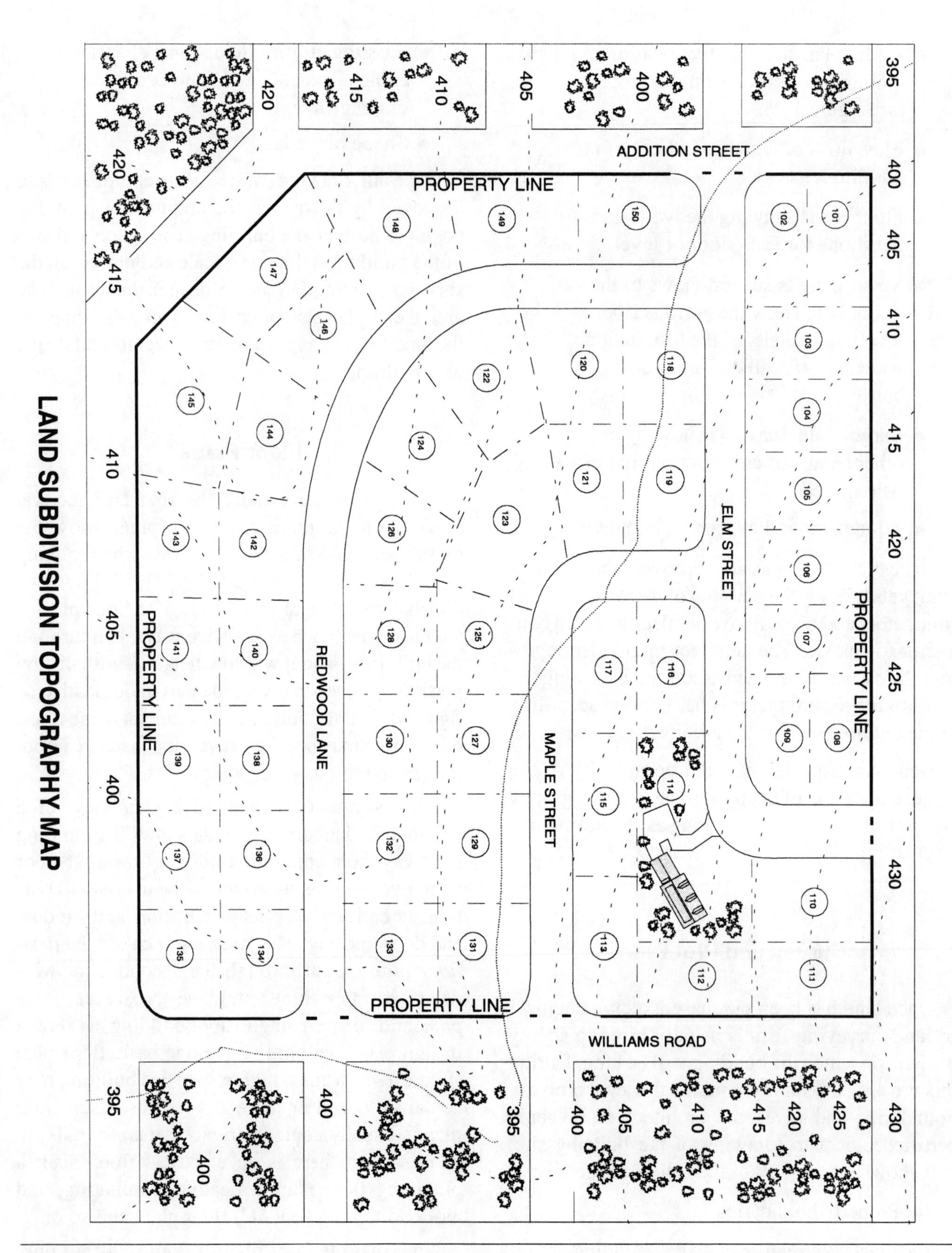

Figure 1-2: Typical plot plan showing the location of the building in relationship to the property boundaries.

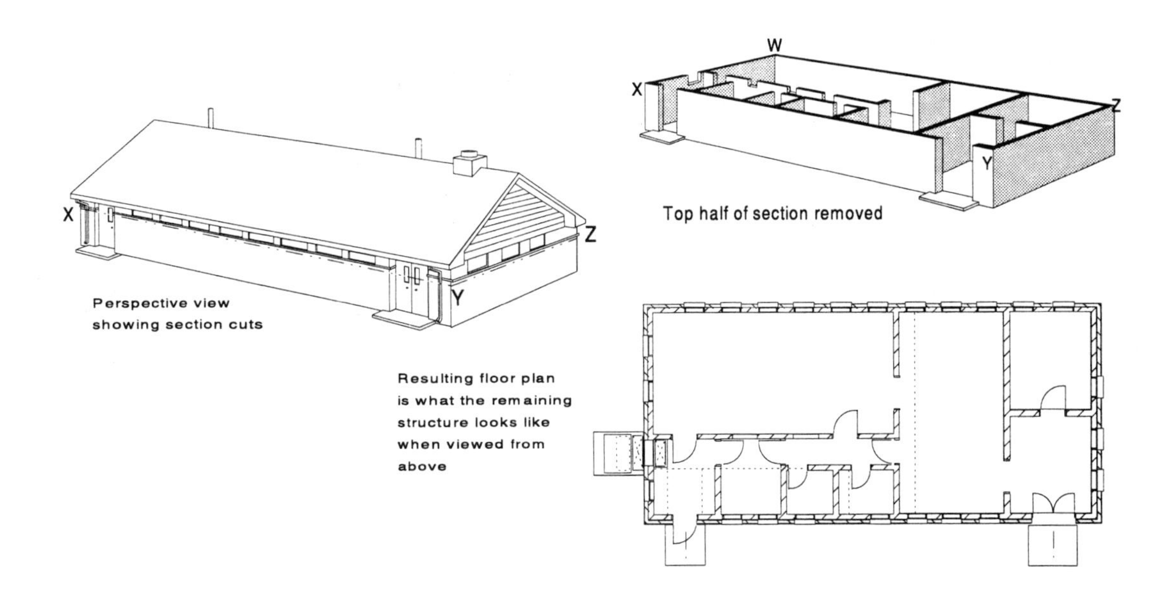

Figure 1-3: Principles of floor-plan layout.

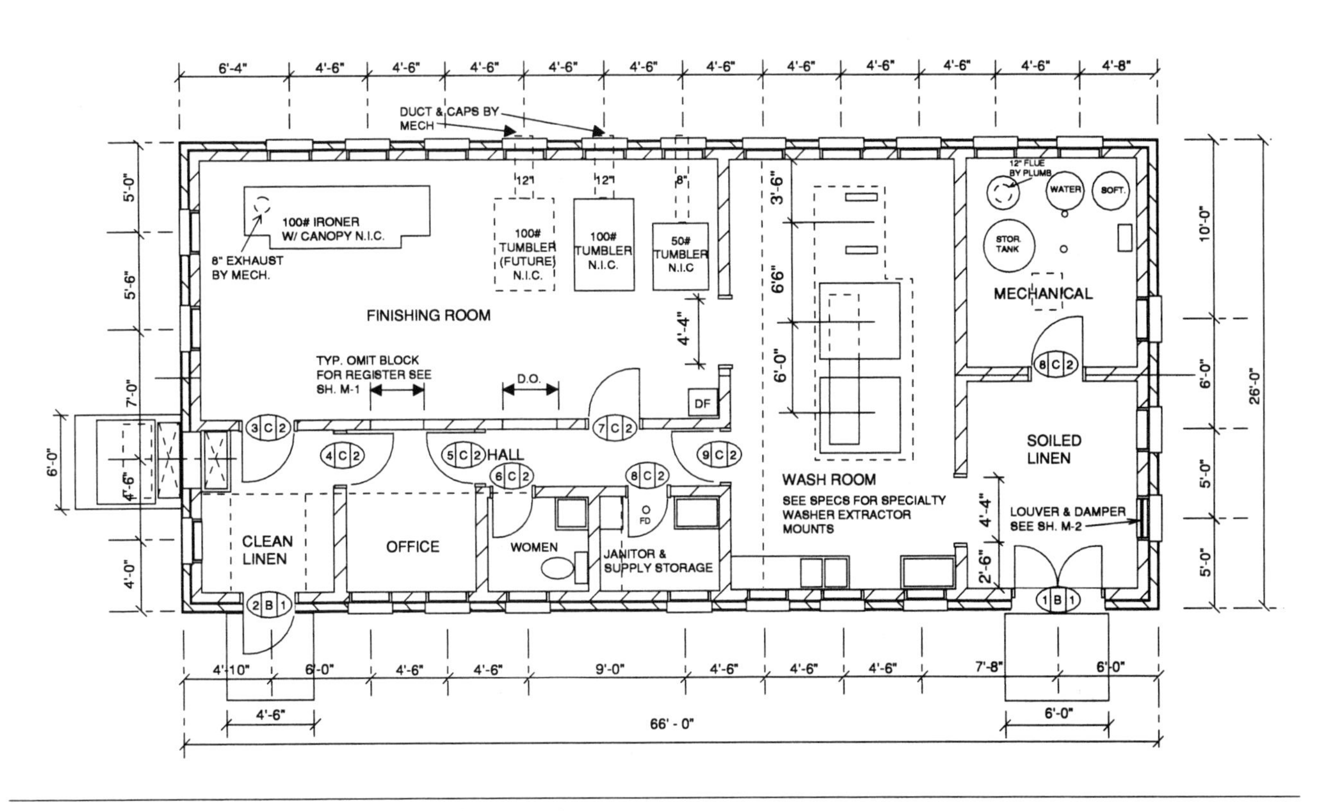

Figure 1-4: Floor plan of a commercial laundry.

doors, but also gives dimensions that tell workers exactly how to construct the building. All dimensions must be checked several times on each drawing to make sure they are accurate.

Once the building is under construction, workers again check all dimensions for accuracy. In fact, workers in all trades must learn to determine if the drawings of the building on which they are working are correct.

Elevations

Once the floor plans are drawn, the architect prepares drawings of the building's elevations. A plan view may represent a flat, curved, or slanting surface. The elevation drawings show which type of surface is involved.

The elevation shows the building's heights and may show the length or width of a particular side. Figure 1-5 shows elevation drawings for our sample laundry building. Note that these drawings include the heights of windows, doors, and porches, as well as the pitch of roofs, and so on—all of which cannot be clearly shown on floor plans. Workers in the various trades need elevation drawings to understand what work they must do and how they must do it.

Sections

A sectional view in architectural drawings is a drawing that shows a building or portion of a building cut through on some imaginary line. This line may be either vertical (straight up and down) or horizontal (straight across). Such views show a building's true shape and interior makeup.

The section line marks the point on the plan or elevation where the imaginary cut has been made. This line is usually a heavy double dot-and-dash line. Arrow points are placed at the ends of section lines to show the direction of the view.

It is often necessary to show more than one section on the same drawing, so different section lines must be identified by different letters or numbers placed at the ends of each line. For example, in figure 1-6, one section line is lettered A; another is called detail section B, and so on. These section letters are generally dark and large so that they stand out on the drawings. The section is named according to these letters—that is, Section A, Detail Section B, and so forth.

A longitudinal section is taken along the length of the building, while a cross-section is usually taken straight across the width. Sometimes, however, a section is taken along a zigzag line to show important parts of the object.

Wall sections are nearly always vertical, so that the cut edge is exposed from top to bottom. The wall section is one of the most important drawings to construction workers because it shows how the structure is built. The floor plans show how each floor is arranged, but the sections tell how each part is constructed and what materials are used. Carpenters, bricklayers, and other workers need this information to construct a building according to the architect's design.

Detail Drawings

Almost all architectural drawings require certain details so that workers understand their particular role in the building's construction. Such information is sometimes given in written specifications, notes, or schedules. However, the most practical way to give this information is to include large-scale detail drawings in the architectural plans. Such drawings may show the actual dimensions of moldings or baseboards, the fitting of a particular wood joint, or perhaps a decorative brick pattern. They are meant to further clarify all the details of the construction project.

Specialty Drawings and Consulting Engineers

All large projects require drawings for the special aspects of the building—structural, electrical, plumbing, and HVAC components. On small projects, the architect may handle these drawings. For example, for the HVAC system, the architect may

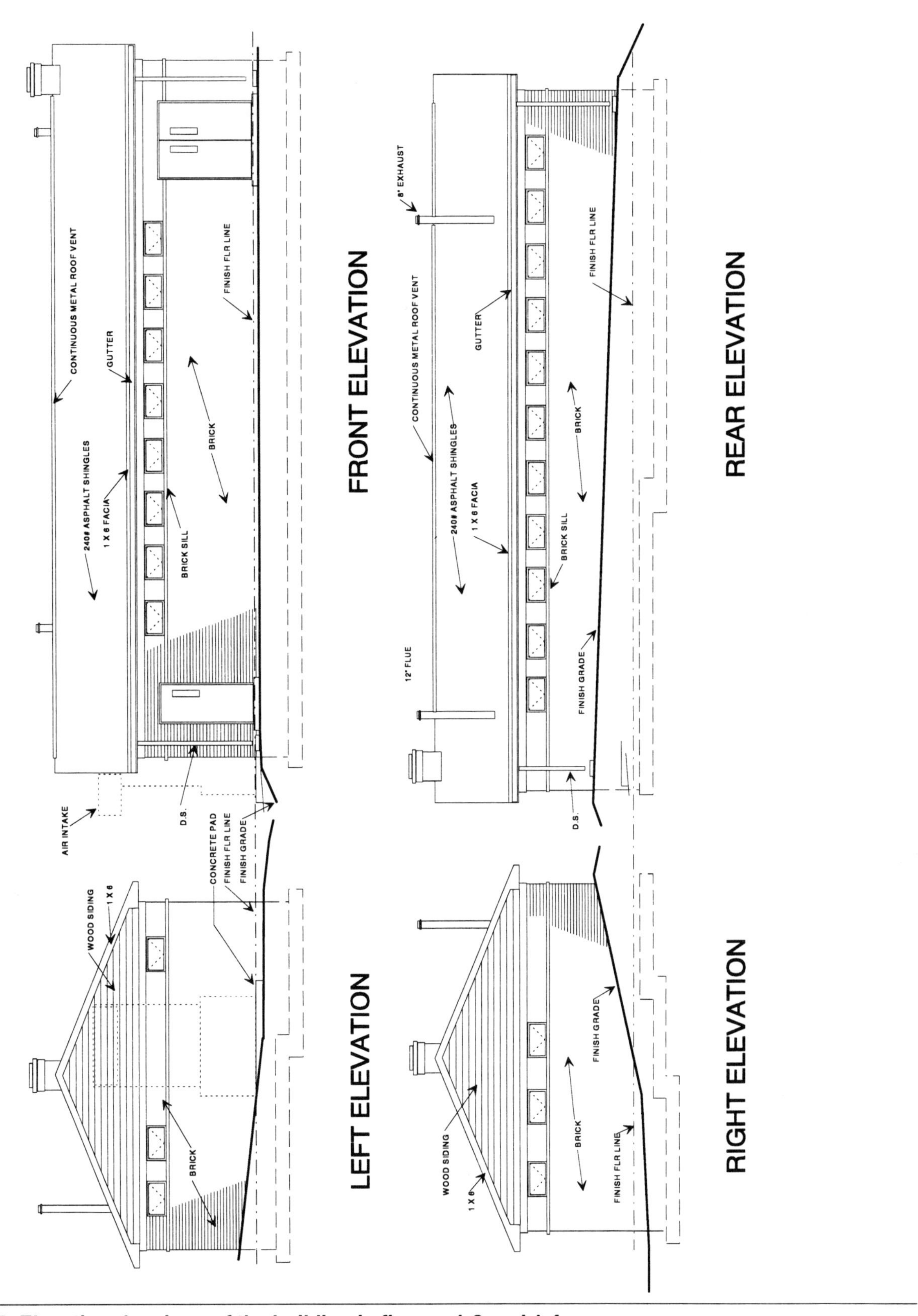

Figure 1-5: Elevation drawings of the building in figures 1-3 and 1-4.

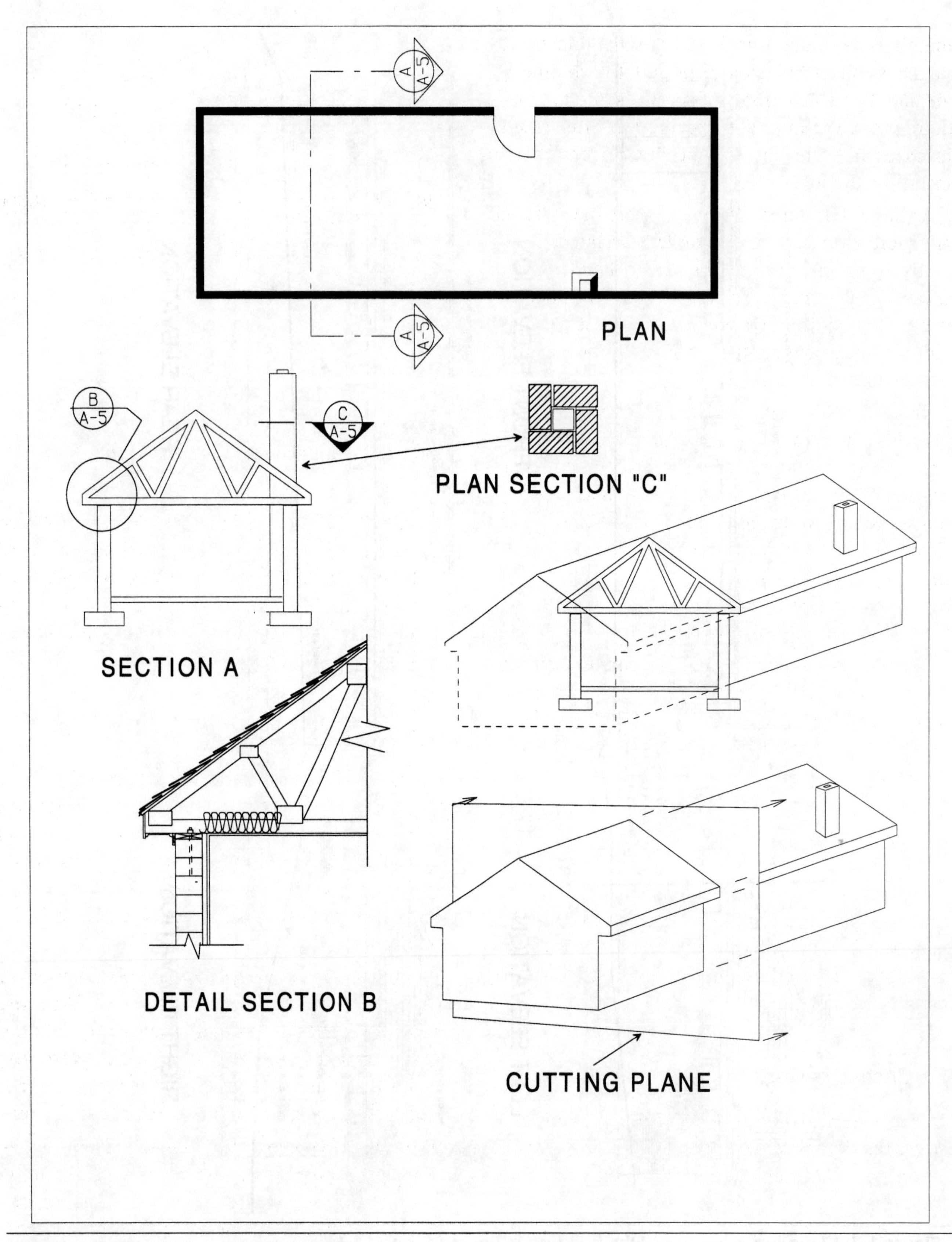

Figure 1-6: Principles of sectional views.

merely show a one-line diagram of the duct work on the architectural floor plans with the volume of air and British thermal units (Btu's) of heat required for each area. The duct work is often sized, in cases like this, by the mechanical contractor installing the job.

In larger buildings, however, where the systems are more extensive and complex, architects normally hire consulting engineers to handle these details. Consulting engineers oversee specialty aspects from building design and layout through bidding and construction to final approval and acceptance of the finished job.

In such cases, consulting engineers often solicit bids for the areas they are responsible for from appropriate contractors. They also inspect their portion of the project to make sure that it is carried out according to the working drawings and specifications. These engineers approve shop drawings (material submittals), check and approve progress payments, and perform other duties that pertain to their phase of the construction.

Architects sometimes ask engineers to estimate the cost of specialty work to help the architect determine the probable cost of the building. Such estimates are especially common in government projects.

Structural Drawings

Figure 1-7 is an example of a typical structural drawing. Structural engineers prepare these drawings based on the weight and stresses the building must bear. These drawings are included with the architectural plans.

Electrical Drawings

The electrical drawings for a building generally cover the complete design of the lighting, power, alarm, and communication systems; special electrical systems; and related electrical equipment. These drawings sometimes include (1) a plot plan or site plan that shows the location of the building on the property and the interconnecting electrical

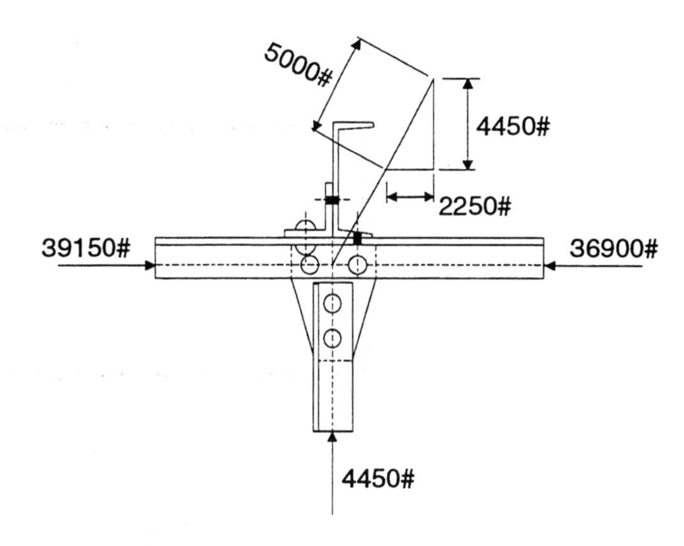

Figure 1-7: Typical structural drawing.

systems; (2) floor plans that show the location of all outlets, lighting fixtures, panel boards, and other components; (3) power-riser diagrams; (4) a symbol list; (5) schematic diagrams; and (6) large-scale detail drawings, if necessary. A typical electrical drawing is shown in figure 1-8.

Mechanical Drawings

Mechanical drawings cover the installation of plumbing and HVAC systems on the premises. These drawings show the complete design and layout of these systems on floor plans, in cross-sections of the building, and in detailed drawings. Wiring for various HVAC controls may also be included.

Figure 1-9 is an example of a plumbing drawing, and figure 1-10 shows a typical HVAC drawing. You are not expected to understand every detail of these various drawings at first glance. However, review them and note every detail to understand the types of construction drawings in use. Experience and further study will clear up any confusion.

Specialty consultants use the architect's drawings for reference when designing suitable systems for a building. For example, the mechanical engineer or designer refers to the architect's drawings when designing an HVAC system. This design

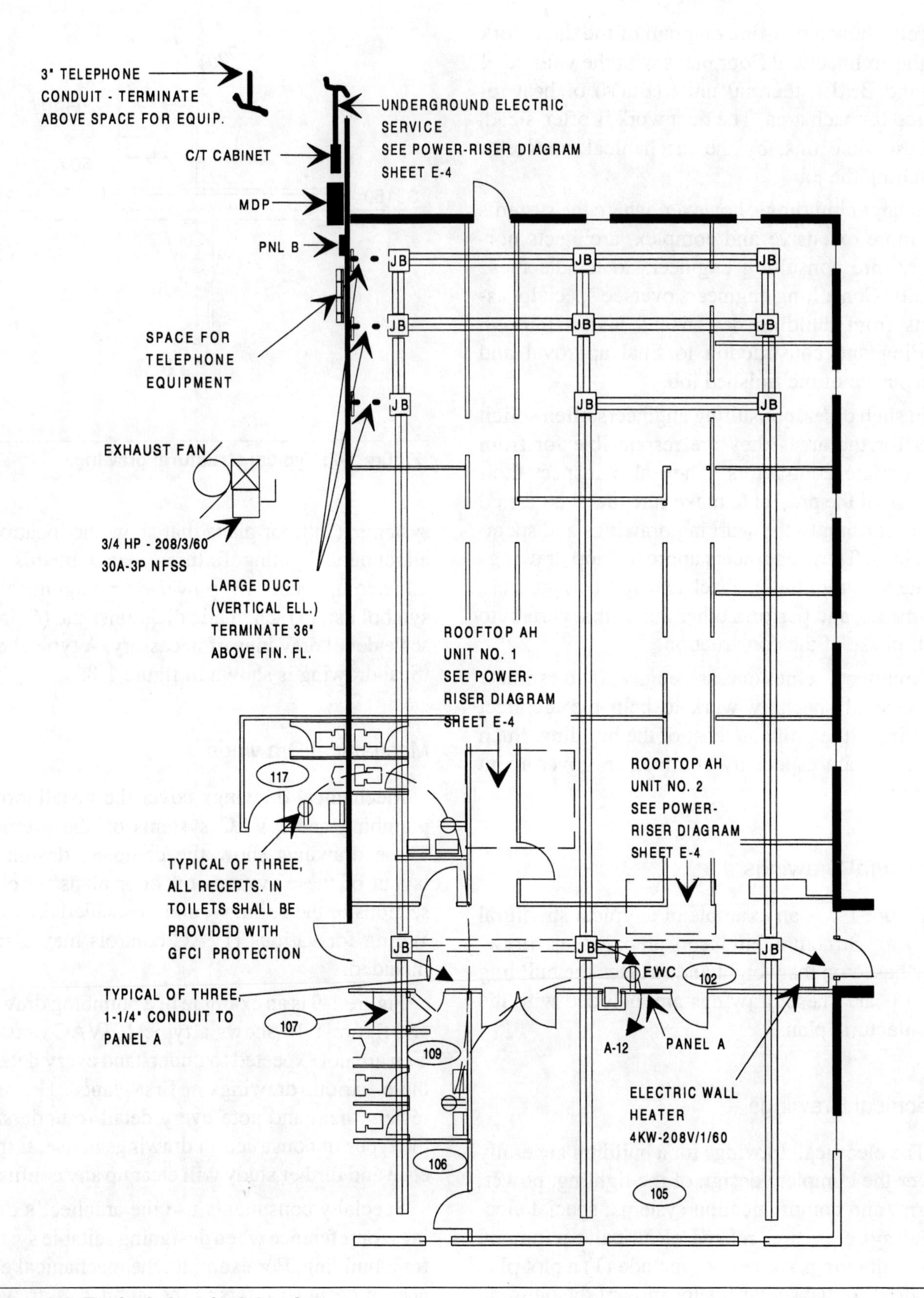

Figure 1-8: Typical electrical drawing.

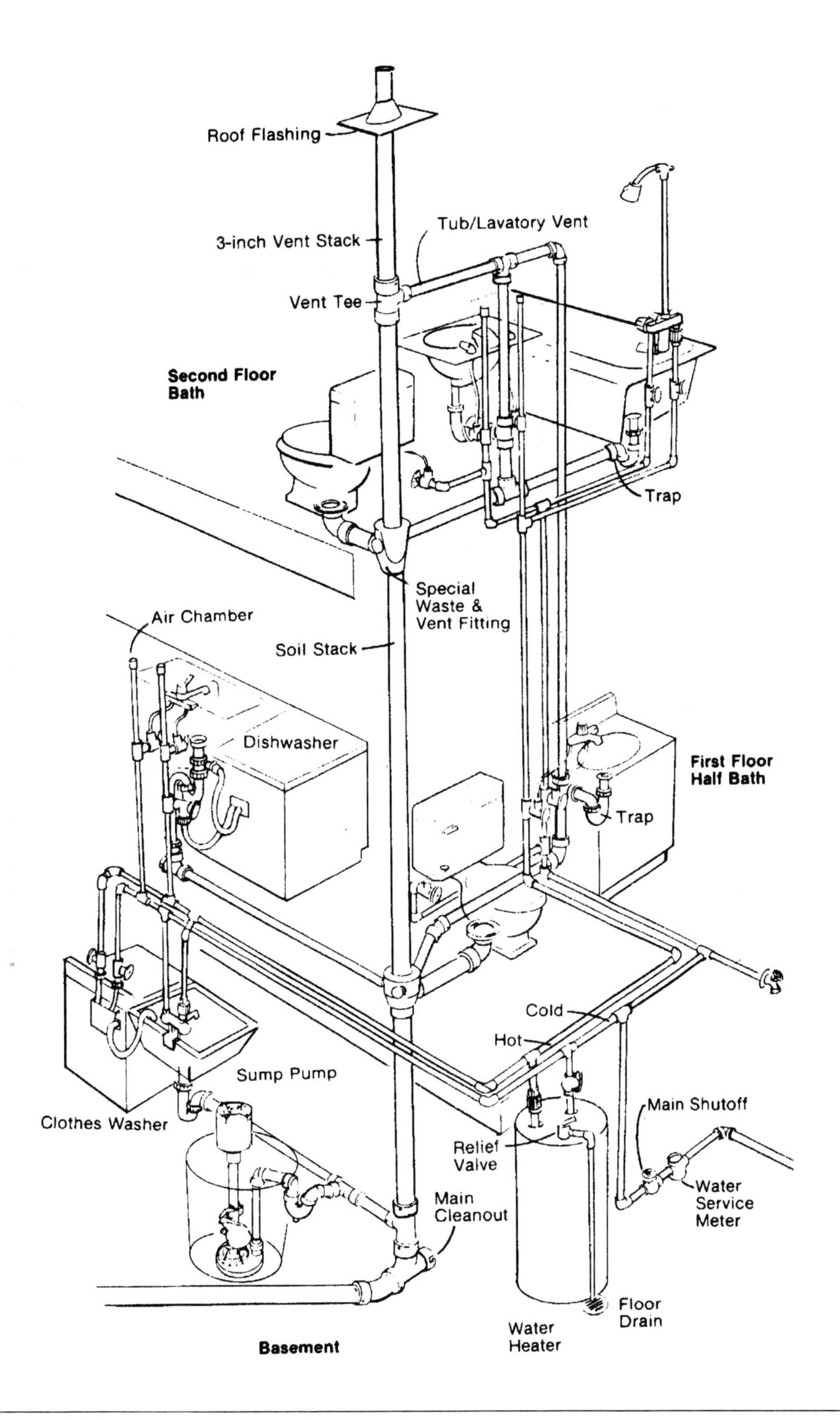

Figure 1-9: Typical plumbing system.

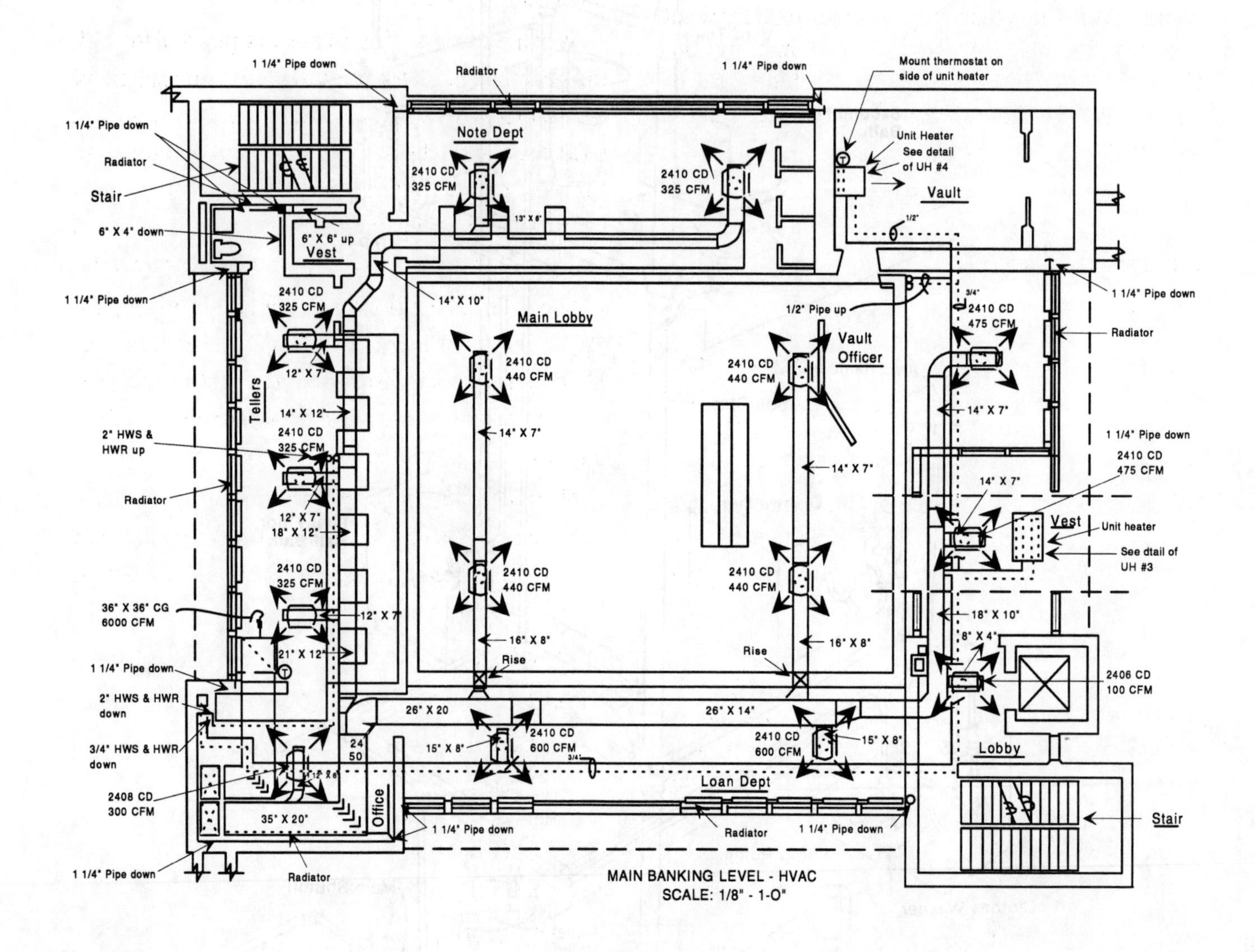

Figure 1-10: Typical HVAC drawing.

usually involves calculating the heat gain for air conditioning, the heat loss for heating, the required air changes per hour for ventilation, the size and shape of ducts needed to deliver the required volume of air to various areas within the building, and similar details. Drafters then translate the engineer's design into neat, detailed, and accurate working drawings that show workers exactly what is required to install the HVAC system correctly. Usually, the architect's drawings include a set of written specifications detailing the requirements of the design.

Written Specifications

Written specifications detail the work and duties required of the owner, the architect, and the engineers. Along with the working drawings, these specifications form the basis of the contract requirements for the construction project.

Once all of these documents are completed, the architect or owner is ready to solicit bids from general contractors so that the construction phase can begin.

Although specialty work is the architect's responsibility, the architect's consulting engineers help solicit bids from electrical and mechanical contractors. Engineers also inspect their portion of the work to assure the architect and owner that this portion is carried out according to the working drawings and specifications.

CONTRACTORS

A contractor is a person or firm that furnishes supplies and performs work at a given price or wage based on a contract. There are many types of contractors—from the small builder or remodeler to huge organizations with offices and projects around the world.

General Contractors

Most construction projects are handled by a *general contractor* who is responsible for the overall construction of the entire project and coordinates the work schedules of all specialty contractors (also called subcontractors).

When a construction project is put out for bid, the general contractor requests bids from specialty contractors—electrical, mechanical, painting, structural, and others. These bids are combined to form the overall bid or price for the entire project. Consequently, most construction projects use both a general contractor and numerous subcontractors.

Some larger contractors maintain tools, equipment, and a staff of trained supervisors and specialty trade workers to complete construction projects. Such an organization does not need to solicit bids from other subcontractors.

Nearly all construction firms have several levels of professionals and workers. These positions are described in detail in Chapter 2, but let's look at the basic ones now.

- Officers of the firm.
- Project managers.
- Safety managers.
- Skilled craft workers.
- Engineers.
- Estimators.
- Drafters.
- Expeditors.
- Inspectors.
- Accounting personnel.
- Purchasing agents.
- Supervisors.
- Foremen.
- Maintenance personnel.

Through training and hard work, chances of advancement are ensured.

IT TAKES THE SKILLED WORKER TO MAKE EVERYTHING HAPPEN!

Much time and expense are involved in any building project before the actual work begins. However, this time and expense would be in vain were it not for the skilled worker. The trades are the backbone, arms, and legs of the construction industry. No building could ever be constructed without them.

It takes excavators to dig the foundation; carpenters to construct forms for the footings; masons, concrete finishers, and ironworkers to construct foundations; masons and carpenters to erect the walls; roofers to finish the roof; electricians to install the electrical system; plumbers to install the plumbing; and HVAC workers to install heating, cooling, and ventilation.

Other specialty trades may also be required, such as millwrights, pipefitters, floor finishers, painters, and instrumentation personnel. It takes a close-working team of skilled professionals to complete a building accurately and on time.

Many opportunities await you if you work hard and can be a team player. First, you must thoroughly learn your trade and then apply this knowledge to the actual job. Books and instructional materials are necessary, but how you use the knowledge is even more important. Great adventure and financial reward await those who succeed in the building trades. Several trade organizations are available to help you get started in almost any area in the United States. Yes, you earn while you learn one of the many fascinating constructions trades—all of which offer good wages and fringe benefits.

EMPLOYMENT OPPORTUNITIES

Many primary trades work in the construction industry (shown in figure 1-11), providing a variety of challenging opportunities.

Other specialized trades in construction include marble setting, tile setting, and equipment engi-

- Carpentry
- Drywall installation
- Electrical work
- Glaziery
- Heating, ventilating, and air-conditioning installation
- Industrial coatings
- Instrumentation
- Ironwork
- Laborers
- Masonry
- Metal building assembly
- Millwright work
- Painting
- Pipefitting
- Plumbing
- Sheet metal work
- Welding

Figure 1-11: Primary construction trades.

neering. Before choosing a trade, consider your personal and professional aspirations as well as the cost of education in relation to the short- and long-term payoffs.

The Bureau of Labor and Statistics (BLS) provides employment and unemployment rates for the construction industry. Additional information about unemployment rates is available from your state labor department.

Depending on your career goals, you can pursue your trade in any of the industry's three sectors:

(1) *Residential*—the construction or renovation of single- and multifamily dwellings, including town houses and tract development homes.

(2) *Commercial*—the construction or renovation of business buildings that can range from small roadside vegetable stands to giant, high-rise office complexes.

(3) *Industrial*—the construction, renovation, or maintenance of industrial plants, factories, refineries, and similar facilities.

Each of these sectors has advantages and disadvantages; but while no one knows what lies ahead for any construction trade, whichever career path chosen is certain to provide a challenging and rewarding future.

Construction workers must be prepared to work in all kinds of weather. However, if you are not suited to such conditions, you may opt for one of the many challenging office jobs that construction firms offer. You must, however, be very knowledgeable about on-site construction to work in these positions. Construction estimators, for example, must have a thorough knowledge of construction processes and materials, but must also be well organized and have reasonably good typing, writing, and math skills. They also use computers to put together their information, so basic skills in computer operation are also necessary.

Basic skills in computer operation are essential for most construction office jobs.

Other office positions in the industry include estimator, construction inspector, safety officer, training specialist, and marketing expert.

Even workers who start out in the field may choose to move to a desk job in later years. If you learn your trade well and have the other skills needed to work in the office, there is likely to be a position for you.

So, if you are interested in building technology and the intricacies of the construction process, but perhaps not attracted to life in the field, consider the many office positions available in construction.

It takes a team of skilled trades to construct any building.

Chapter 2

Organization and Structure

The organization and structure of the construction industry have changed dramatically in the past century. Recent improvements in technology have enabled the industry to thrive.

The construction industry is made up of numerous entities (figure 2-1). Its purpose is to provide construction services to the public and private sectors using highly advanced equipment and materials. Its backbone is its work force, which must be well trained. Numerous education and training programs in the trades are offered today by vocational schools, high schools, and colleges. Apprenticeship programs are also available.

The industry is organized into three sectors—residential, commercial, and industrial construction. The residential sector supports construction of single- and multifamily dwellings. Projects range from home alterations and improvements to the construction of modular homes, land-tract developments, and town houses. The U.S. Department of Commerce has forecasted that by the year 1997, $182 billion will be spent on residential construction projects, of which $113 billion will go toward single-family dwellings, $15 billion toward multifamily dwellings, and another $54 billion toward home improvements and alterations.

The commercial sector supports construction of nonresidential buildings including stores, churches, schools, high-rise offices, and shopping complexes. The Department of Commerce forecasts that this sector will spend $109 billion on nonresidential construction in 1997. Another $109 billion will be spent on public works and industrial projects.

Within each sector are numerous contractors who provide the end products—office buildings, homes, bridges, highways, and so forth.

Contractors are also organized into three types: (1) general and operative building contractors, (2) highway construction contractors, and (3) special trade contractors (figure 2-2).

All three groups are vital to the industry, and each provides advancement opportunities to

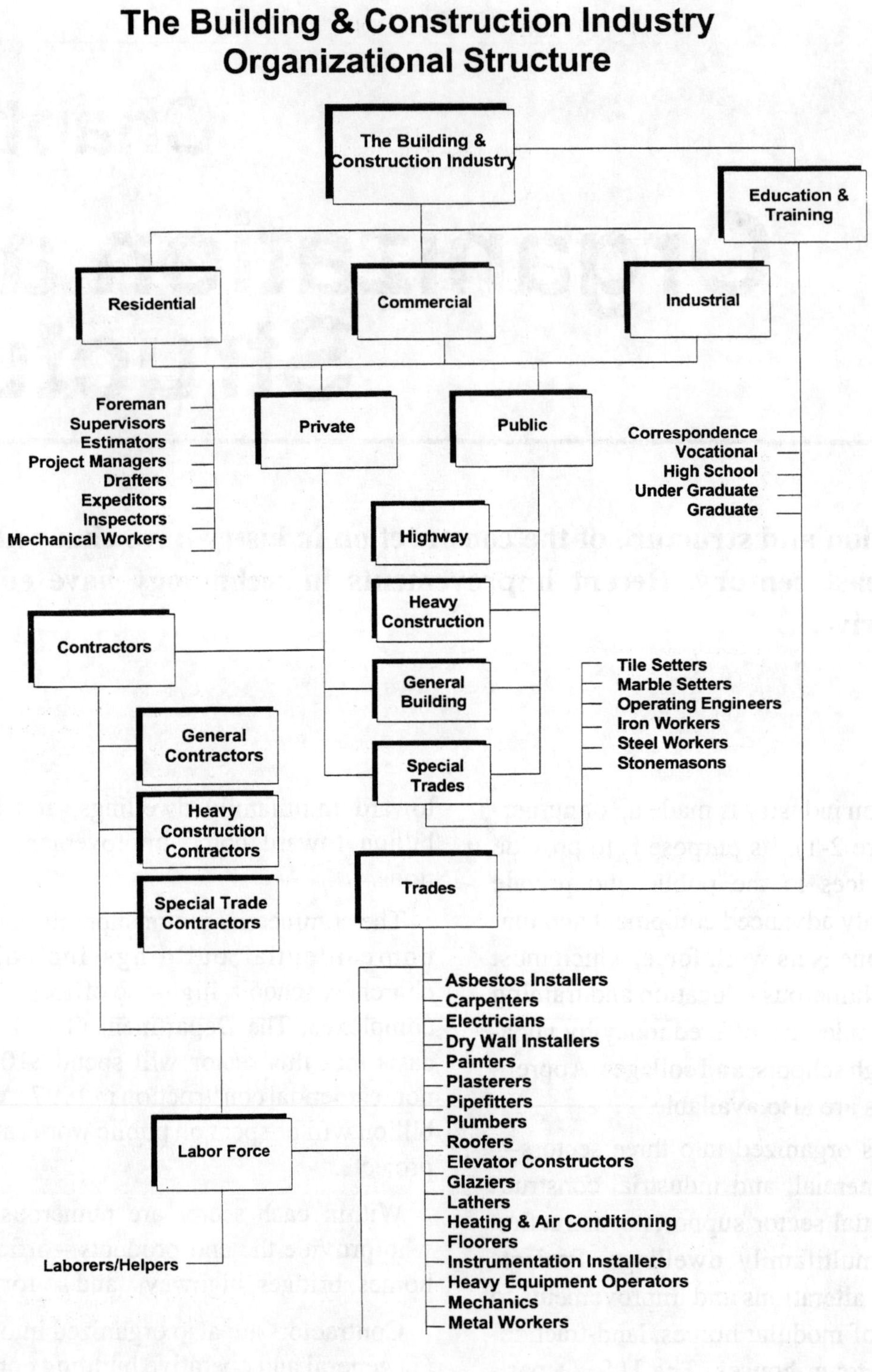

The Building & Construction Industry Organizational Structure

The Building & Construction Industry

Education & Training

Residential

Commercial

Industrial

Foreman
Supervisors
Estimators
Project Managers
Drafters
Expeditors
Inspectors
Mechanical Workers

Private

Public

Correspondence
Vocational
High School
Under Graduate
Graduate

Highway

Heavy Construction

Contractors

General Building

Special Trades

Tile Setters
Marble Setters
Operating Engineers
Iron Workers
Steel Workers
Stonemasons

General Contractors

Heavy Construction Contractors

Special Trade Contractors

Trades

Asbestos Installers
Carpenters
Electricians
Dry Wall Installers
Painters
Plasterers
Pipefitters
Plumbers
Roofers
Elevator Constructors
Glaziers
Lathers
Heating & Air Conditioning
Floorers
Instrumentation Installers
Heavy Equipment Operators
Mechanics
Metal Workers

Labor Force

Laborers/Helpers

Figure 2-1: Organizational structure of the building and construction industry.

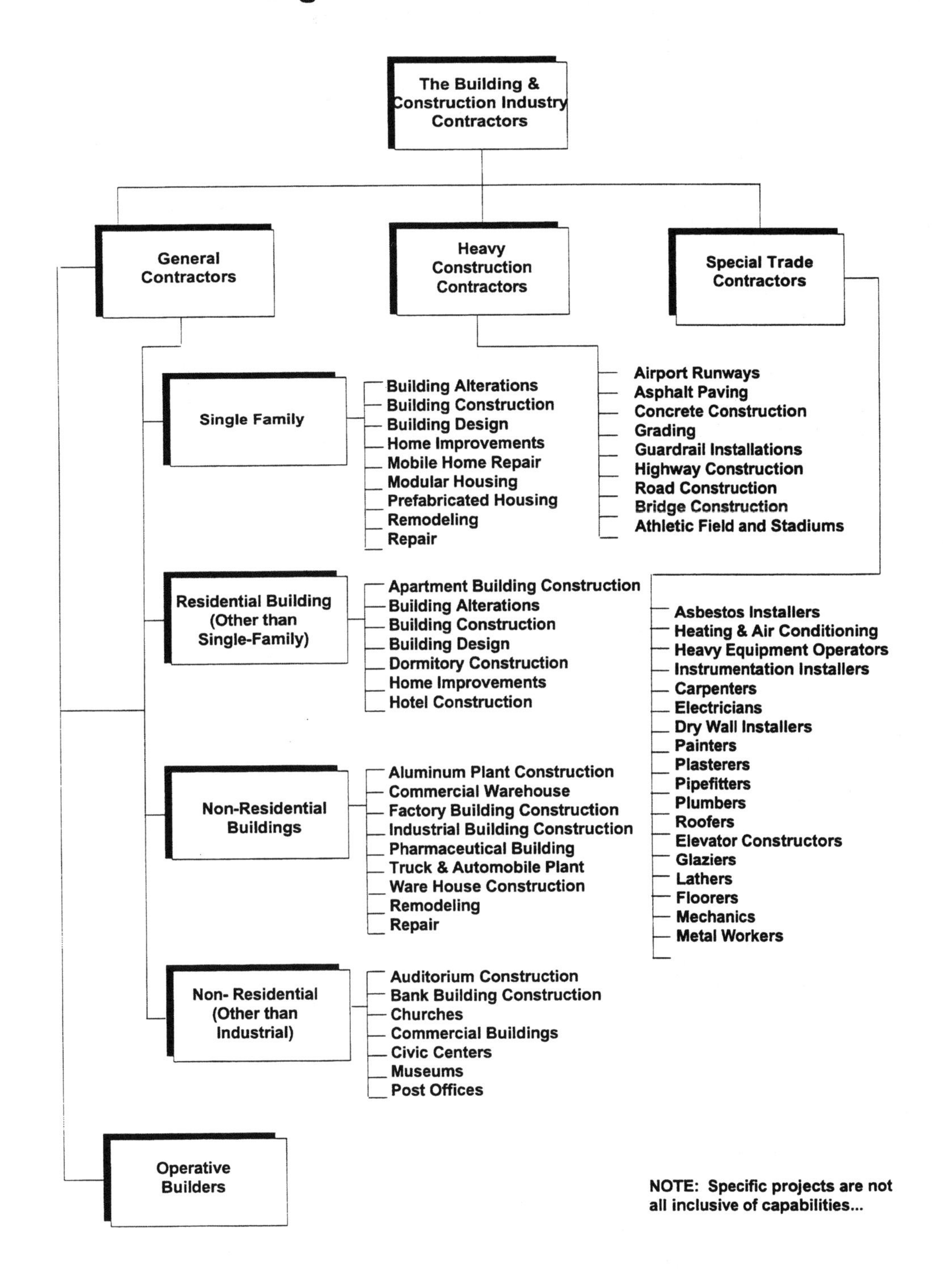

Figure 2-2: Organizational structure of contractors in the building and construction industry.

skilled workers. Many trade workers became contractors after years of education and on-the-job training. They advance to positions as foremen, supervisors, and project managers, and thus gain the skills required to become a contractor.

GENERAL CONTRACTOR

Residential

Residential general contractors are entrepreneurs within the construction industry. Using their own finances, these contractors frequently purchase and develop land sites for future homes. The general contractor specializing in residential construction is engaged in all facets of construction projects and is mainly concerned with constructing single- and multifamily housing units. Projects include the construction of new homes, additions to older homes, alterations, remodeling, and repairs. Residential general contractors are usually not equipped to construct entire buildings on their own. They usually have their own carpenters on staff, as well as a few other trade workers. But when the project requires many workers, these builders subcontract some or all of the following work:

- Drywall installation.
- Electrical wiring.
- HVAC installation.
- Ironwork (decorative).
- Masonry.
- Painting.
- Plumbing.

Operative Builder

All contractors are subject to some financial risk, but the operative builder's risks are the greatest. Unlike the general contractor, the operative builder constructs numerous residential and nonresidential projects on speculation—that is, without having a contract to do so. Operative builders handle new home construction, alterations, remodeling, and repair. They carefully research their projects; buy and sell property (land, existing residential and nonresidential homes, etc.) and build on the land or renovate a building in the hope of realizing a profit. In addition to the usual aspects of construction, the operative builder must have a knack for real estate appraising, mortgage foreclosures, and pricing materials and equipment.

Industrial

The industrial general contractor normally constructs warehouses, industrial complexes, automobile assembly plants, pharmaceutical manufacturing plants, oil refineries, and so on. Most industrial general contractors finance their own projects and usually subcontract specific portions of a given job. These contractors normally need more equipment, personnel, and finances because of the size and types of projects they undertake. For example, a pharmaceutical building may require certain materials and mechanical instruments that are not often used, so the general industrial contractor may have to subcontract with an instrumentation specialist.

Most large industrial contractors keep key workers from each trade on staff. When a new project begins, these workers supervise their particular trade at the job site. The trade workers are usually hired locally.

Commercial

General contractors who specialize in commercial work are usually organized like industrial contractors, except on a smaller scale.

Commercial general contractors build shopping malls, office buildings, apartment buildings, stores, and similar structures. In some areas, churches, libraries, schools, hospitals, and college buildings are considered commercial work; in other localities, these projects are considered institutional.

Heavy Construction Contractors

Heavy construction contractors are general contractors who specialize in heavy construction projects such as highways, streets, railroads, sewer systems, airport runways, bridges, tunnels, and elevated highways. These contractors fall into two major groups: (1) highway and street construction and (2) all other heavy construction projects. The first group builds highways, roads, streets, sidewalks, airport runways, and similar projects. The second group builds such structures as bridges, dams, tunnels, and sewer systems.

Special Trade Contractors

Special trade contractors primarily work on the construction of homes, commercial complexes, and industrial plants. The specialized trades include plumbing, heating and air conditioning, painting, electrical work, and masonry; and one contractor may specialize in several areas. Specialized trade contractors usually work as subcontractors to a general contractor. The general contractor subcontracts work when it is cost-effective to do so or when the general contractor does not have the required expertise. Specialized trade workers are called on throughout different phases of the construction project. Normally, the general contractor subcontracts with specialized trade workers before construction begins so that they can order supplies and obtain necessary equipment, materials, and workers.

As subcontractors, specialized trade contractors report directly to the general contractor and must complete their work within a specified time. The general contractor is responsible for the entire project, including the work done by all subcontractors. The subcontractor is only responsible for a specific segment of the project. Subcontracted work ranges from clearing a lot to installing electrical cable and wiring. Subcontractors often work on several projects at once and may need to hire other skilled trade workers and laborers in order to complete the projects on time.

PRIVATE AND PUBLIC CONSTRUCTION

Private Construction

Private construction refers to the construction of residential, nonresidential, and farm buildings, and accounts for most of the construction industry's work force. Half of all private sector construction is residential, 15 percent is commercial, and 5 percent is industrial. The remaining 30 percent is alterations, remodeling, and construction of public utilities.

Public Construction

Public construction refers to the construction of highways, streets, bridges, and so on. The public construction sector has close ties with federal, state, and local government. Other public construction projects include education buildings, hospitals, and public housing units. Residential construction (i.e., public housing) accounts for 2 to 3 percent of public construction. The remainder is construction or modification of public highways, bridges, roads, utilities, and similar projects.

COMPANY POLICIES AND PROCEDURES

Key to any construction organization are its company policies and procedures, which provide the means for carrying out management processes, aid in decisionmaking and problemsolving, and provide consistency to business operation.

Policies and procedures vary widely from one contractor to the next; but most fall under the following general headings:

- Payroll practices
- Safety
- Drug and alcohol use
- Personnel

Construction Projects

Single Units 1980-1993

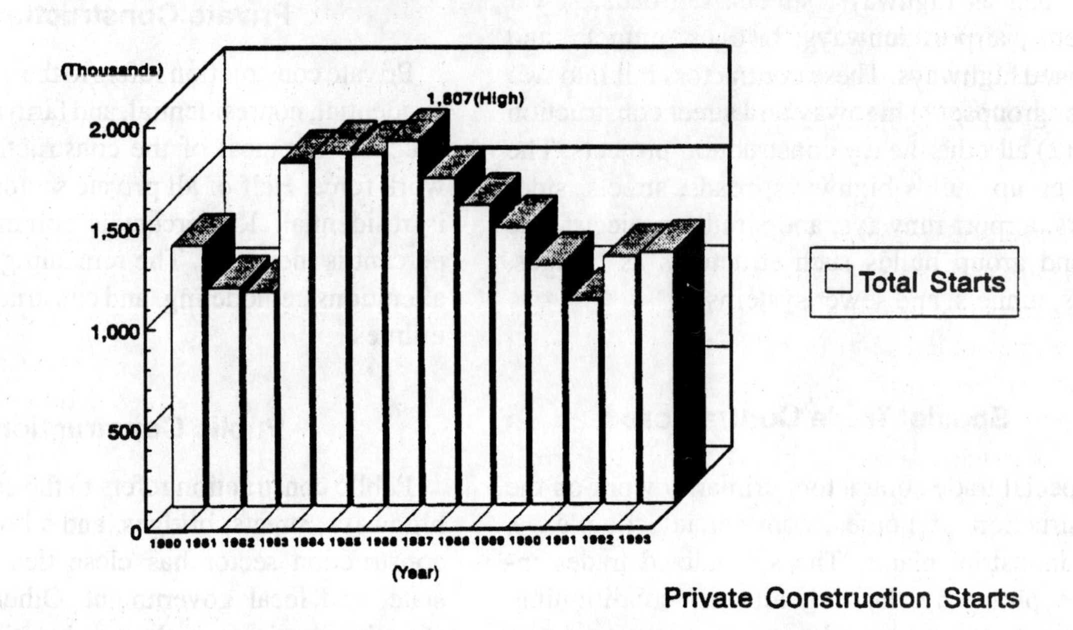

Figure 2-3: Percentage of construction projects from 1980 to 1993.

- Work schedule
- Benefits

Company procedures are designed to improve overall efficiency, to reduce the cost of running a business, to standardize the ways in which work is performed, and to create a smooth work atmosphere.

Some smaller contractors may give information about company procedures orally, but most contractors provide this information in writing. Sources of this information include the following:

1. **Company Manual.** This publication provides general information about the company. It may describe the history and function of the company and its organizational structure. It may also give the names of company and departmental or division executives.

2. **Employee Manual.** This publication may be separate from the company manual or included in it. In some cases, employee information may simply be posted on a bulletin board.

   This manual provides information or policies regarding work schedules, vacations, sick leave, holidays, company benefits, and so on. Some firms provide a separate manual for each job or trade.

3. **Office Manual.** This publication describes how specific job duties are performed. It may also include information about procedures for typing correspondence, billing, paying bills, ordering materials and equipment, answering the telephone, receiving guests, and similar office activities.

4. **Job Manual.** This publication usually describes each job, the firm's expectations of employees, and the supervisor for each job category.

ENTERING THE CONTRACTING BUSINESS

Entry-level trade workers should always have their eyes open for advancement possibilities. Set your goals early and work toward them, even though these goals may change throughout your career.

Enthusiastic workers who are properly trained and have a good work attitude will advance rapidly from apprentice or trainee to first-class technician (sometimes called a *journey worker*), then perhaps to foreman of a work crew and supervisor of an entire project. The possibilities are endless. Some trade workers may consider opening their own business. But even if you are not interested in running your own firm, understanding the highlights and pitfalls of contracting will help you do your job better.

Contracting can be profitable, but there are many hazards to getting started. The sooner you realize the hazards, the sooner you will be able to operate your business smoothly and safely.

If you are considering such a venture, first take a long, hard look at yourself to determine if you are in fact qualified. Are you satisfying your current employer? If not, you would probably have a hard time satisfying customers in the contracting business. On the other hand, if you are well trained, have a good knowledge of building codes, are familiar with modern installation techniques, and can direct people well, you are off to a good start.

However, being skilled in construction work doesn't guarantee success in contracting. For example, you may be an excellent technician, foreman, supervisor, or estimator, but if you lack sufficient financing, executive ability, or general knowledge of the business, you are not qualified to operate a contracting business.

Other important requirements are confidence and enthusiasm. Confidence in yourself and enthusiasm about overcoming obstacles add to your chances of success.

You should also have a good knowledge of specialty codes (for example, the National Electrical Code and the National Plumbing Code), a relatively good knowledge of design, reasonable mathematical skills, the ability to accomplish goals *through* your workers (not by workers), the willingness to make sacrifices, and a good knowledge of estimating and business fundamentals.

Successful contractors practice extremely efficient management techniques; use only workers specifically trained in, experienced in, or adaptable to the work performed; and develop an accurate and fast system of estimating the cost of jobs.

If you are weak in any of these areas, there are courses available at colleges, junior colleges, vocational schools, and other institutes on all phases of the construction business. There are also many good books and self-study courses.

Equipment

How much office space and equipment will you need for your new business? What kinds of tools should you buy, and how many of them should you have? The answers to these questions depend on the type of contracting work you go into and the volume of business you anticipate. Experienced contractors recommend that you start with a minimum amount of tools and buy new ones as needed. A contractor who invests heavily in tools can run into problems. For example, if a small contractor invests $140,000 in tools and equipment, and the work slacks off or stops, a lot of money is tied up in a tool inventory that is worthless at the moment.

Business Location and Layout

A typical office may occupy a single room, an entire floor, or even a whole building. Factors that dictate the size of the contractor's office include volume of business and the availability of suitable space.

The physical arrangement of a contractor's shop and office affects how efficiently a job gets done and how effectively materials and tools are used.

If the building you occupy has limitations on floor area, accessibility, relation to the street, yard area, and so on, it may not be possible to have an ideal physical layout; however, if you plan intelligently, you can usually set up operations that run smoothly and efficiently.

A good physical arrangement allows for the following:

- Easy and minimal handling of materials and tools.
- Accessibility.
- Sufficient storage areas for assembled orders and for materials and tools returned from jobs.
- Unobstructed truck access.
- A centrally located stock room desk or office.

These objectives are best accomplished when the following conditions exist:

- All facilities are located on one floor.
- Adequate floor space is available.
- Sufficient bins, shelving, and storage racks are available and separated by adequate aisles and walking space.
- Separate receiving and shipping areas or driveways are available. (Driveway access should be provided from streets that do not have a lot of traffic.)
- Platforms high enough for truck beds are available for unloading, storing, and loading equipment.
- Yard space is accessible to trucks for parking, storage, receiving, and shipping—preferably with receiving and shipping entrances opening from the building directly to the yard.

The entire arrangement should follow a production line as closely as possible, so that each area of the shop prepares the project for the work to be done at the next area. Related materials that are shipped at the same time should be grouped together. Workers with the same skills should work as a team.

When available floor space is restricted, balconies can be constructed and used for storing items that are handled and shipped less frequently.

Office Personnel

Key personnel in a contracting firm include executives, administrators, estimators, salespeople, and sometimes diplomats. All of these people must have a good working knowledge of building codes and ordinances and of construction in general.

Estimators are especially important. It is their job to determine a reasonably accurate price for a project. That price must provide sufficient income to cover all material and labor costs, direct job expenses, and overhead or operating expenses, and still leave a margin of profit. Bad estimators can break a firm in a very short period of time.

Consequently, estimators are among the highest-ranking members of any contracting firm and can command top salaries. If you are qualified to be an estimator, this position is one to set your sights on.

Job Supervisors

While most owners and principals of contracting firms are experts in their respective fields, they must select the best supervisors, foremen, and other employees to run a given project. Just because supervisors can oversee one type of project does not mean that they can handle all types.

Foremen or supervisors who have supervised mostly commercial projects will need time to adjust to industrial or large institutional projects. Supervisors who have worked mostly on larger projects will need time to adjust to smaller ones.

Foremen and supervisors must be skilled at directing others under the circumstances of the project. A foreman who can direct a small crew of workers well may be unable to control a large crew.

Supervisors normally have direct contact with architects, engineers, and owners of buildings. Therefore, they must be tactful and have good "people skills." They must be able to control a job rigidly without being unpleasant.

Executives of contracting firms should look for supervisors and foremen among their existing employees as well as elsewhere.

Contractors who recognize the ability of their employees often build an excellent work force. Many workers are good technicians but poor leaders. Others like to take the lead and assume responsibility. Some people have better temperaments than others for supervising employees, but in general, supervisors should be able to control the work rigidly without having to be unpleasant. The majority of supervisors hired by contractors are high-caliber people. They want to feel that they are working *with* employers rather than for them. Employees should always be assigned to the work they are best suited for. Contractors should consider all of the characteristics of their employees and try to work out the best arrangements for each.

Contractors should never try to build a good work force by carelessly hiring and firing employees. This practice is costly and harmful to a firm's reputation. Constant hiring and firing ruins the attitude of steady employees and may even cause a firm to lose customers.

Every effort must be made to select the right people during the first hiring. If such workers are not available, select employees who are willing to learn and can readily adapt themselves to the type of work you expect to contract.

You can learn much of what you need to know about potential employees by interviewing them. Ask employees about their educational background, what type of projects they have worked on—for whom and for how long. Contact previous employers if you are unsure about the worker's honesty and ability.

Once you have built a good organization, you must keep it intact. Treat employees—and direct their work—in a way that will make you a desirable employer. Be courteous and supportive. Each employee is an individual and must be treated with respect.

Chapter 3
Work Habits

The first six months of on-the-job training are the most important in a construction worker's career. During this time, supervisors carefully evaluate work habits, probably more closely than at any other time during training. A supervisor's report from this period can help or haunt the worker's career for years to come.

STARTING A NEW JOB

Your attitude toward your employer coworkers, and the job itself is critical to your job success. Others constantly observe both your disposition and your dedication. In fact, the way you act around others reveals your outlook on life. Set realistic goals and work toward them—always with a good attitude. Many new workers either lose their job or are bypassed for promotions because of a poor attitude.

The Employee

Employers look for certain characteristics in their employees, regardless of the job. These include the following:

> *NOTE*
>
> A worker's responsibility involves more than just performing a job. Employers expect workers to work with a positive attitude toward the job and other employees.

- *Cooperation*—sharing tasks when appropriate.
- *Honesty*—being truthful with employers.
- *Initiative*—showing interest in developing new job skills.
- *Willingness to learn*—learning everything possible about the job.
- *Dependability*—being on time for work and having few absences.
- *Enthusiasm*—showing a sincere interest in job performance.
- *Acceptance of criticism*—using criticism as positive input and trying to improve job performance as a result.
- *Loyalty*—not complaining even if you disagree with a company rule or policy, and keeping company business confidential.

- *Punctuality*—being on time for work and taking only the time allowed for breaks.
- *Regular attendance*—being absent only when ill or in case of an emergency, and notifying your employer so that you can be replaced temporarily.
- *Performance*—always doing your best.
- *Respect for rules*—always obeying company rules and regulations.
- *Proper dress*—wearing clothes suitable for the type of job.

There are also certain ethical characteristics that every employee should have; for example, always providing an accurate account of how your work time is spent, working even when your supervisor is not present, and respecting the firm's tools and equipment.

Job Interviews

Entry-level construction trainees are not expected to have all the skills required for a job without supervision. Training involves learning those skills on the job. However, new job applicants should know the name and description of the job for which they are applying *before* going on the interview. (Chapter 5 describes most construction trades.) Learn unfamiliar words or terms ahead of time. You can find such words in a technical dictionary—know their meaning and spelling. This knowledge will help you fill out job applications.

You should also know some of the tools and equipment that are used on the job. The public library has many books on construction subjects. Use them to get a good overview of the industry and individual trades before your interview. Figure 3-1 gives other guidelines for interviewing.

- Be on time.
- Look your best.
- Be courteous.
- Avoid out-of-place humor.
- Be friendly.
- Be willing to learn.
- Be honest with your answers.
- Sit in a straight, but relaxed, dignified manner.
- Ask questions if certain procedures are not clear.
- Look at the person who is talking.
- Write down information that seems important and worth remembering.

Figure 3-1: Interview guidelines.

EMPLOYEE EXPECTATIONS

What should you expect from your employers?

- *Salary*—that you will be paid for the work you do.
- *Safe working conditions*—that you will have reasonably safe working conditions. Federal laws forbid people under the age of 18 to work in certain dangerous jobs.
- *Training*—that you will receive the necessary on-the-job training.
- *Introduction*—that you will be introduced to all employees with whom you will be working.
- *Rules*—that your employer will explain in detail all business policies, rules, and regulations.
- *Changes*—that your employer will explain any changes that may be made to your job or position, such as in duties, responsibilities, salary, work schedule.

- *Work evaluation*—that your employer will periodically evaluate your job performance.
- *Workers Compensation*—that your employer will provide some form of insurance for injuries received while on the job.

Lost Time

Employers hire the least number of workers required to complete their projects. Consequently, workers must pull their own weight to get a job done.

Employees who miss time burden other workers and slow the project. Consequently, lost time should be kept to the bare minimum.

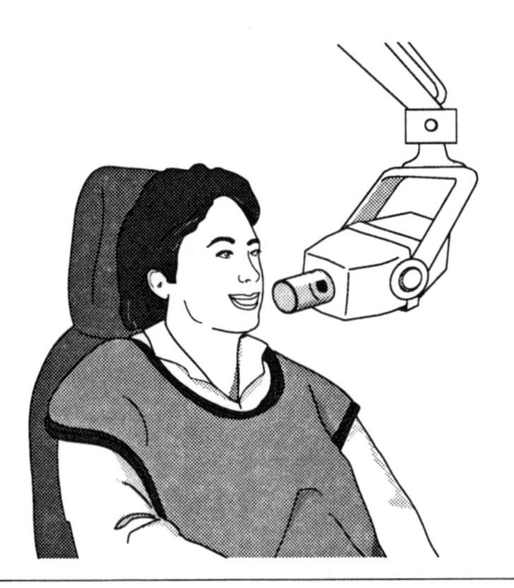

Schedule all dental and other medical checkups on a non-work day.

Everyone experiences illness or other problems occasionally, and some missed work is unavoidable. If you are sick, tell your supervisor or employer and give all the details of the illness. If you see a doctor, get a report from the doctor about the illness and the length of time that you will be off. Such information helps your employer work around your absence. Try to schedule dental and other medical checkups, as well as any personal business, for your days off or after work hours.

Don't party or stay up late during the workweek. Your employer deserves an employee who is on time and alert.

If you become ill, notify your supervisor, giving a description of your illness.

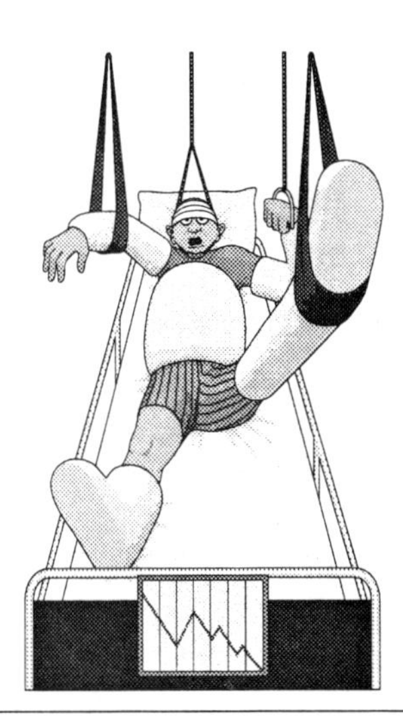

If your illness is prolonged, try to get a doctor's report stating the length of time you will be absent.

Tardiness

Being late for work is as bad as being absent. You might occasionally get stuck in traffic; your car might break down; or a tire might go flat. But these instances should be rare. It is your responsibility to get to work on time. Make sure that your

A mechanical breakdown may be unavoidable, but such mishaps should not happen too often.

car is reliable and in good repair. If necessary, use car pools or public transportation. If your car pool is late, have a backup plan. If your home is a long way from the job site, rent a room closer to the job at least during the week. If something happens to lengthen your commute—road construction, for example—leave earlier.

If you stop for breakfast on your way to the job site, choose a restaurant that is close to work so that the bulk of the trip is behind you.

If you frequently oversleep, try to get more rest at night; avoid late-night activities during the workweek.

Alcohol and Other Drugs

Using alcohol or other drugs during work hours is usually grounds for immediate dismissal from any construction job. The use of drugs affects your performance and endangers the safety of other workers.

If you must take prescription medication for an illness, tell your supervisor. Depending on the type of work you do, your supervisor may assign you to a different position or give you a short leave of absence until you recover.

Many employees believe that what they do off the job is strictly their business. However, if what you do off the job affects your ability to perform on the job, then it *does* concern the employer. For example, if you come to work on Monday morning with a hangover and cannot perform your required tasks well, your weekend activities become the concern of your supervisor or employer. If such occurrences happen often, you will probably be one of the first to go when layoffs occur or you may be overlooked for promotion.

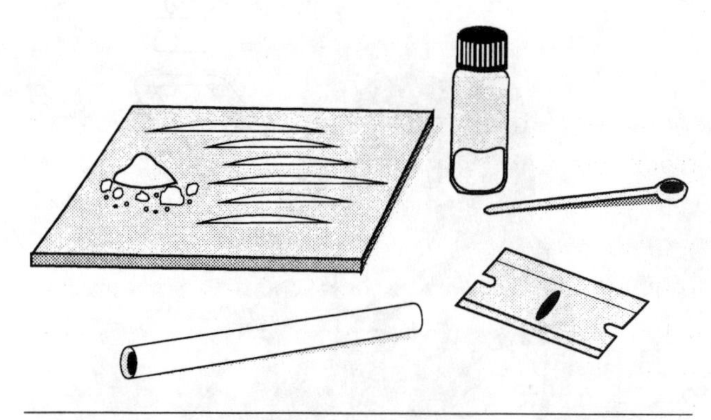

The use of any type of drugs is grounds for immediate dismissal.

If you have a drinking or drug problem, there are many places to get help. Your employer may even sponsor a program.

Because of the hazards involved in construction, this industry has no room for drug users or problem drinkers. Curtail the problem or you will soon be seeking another occupation.

If your off-the-job activities affect your on-the-job performance, then your off-the-job activities become the concern of your supervisor and employer.

Practical Jokes

Practical jokes and "horseplay" may have their place, but not on construction projects. To maintain safety and to perform work efficiently, every worker must be serious and alert at all times. An innocent practical joke may cause injury on the job. For example, a toy rubber snake was placed in the lunch box of an electrical worker who was extremely frightened of snakes. He opened his lunch box on a 30-foot scaffold and when he saw the snake he fell and was injured. As a result, he could no longer work in construction.

Horseplay should be reserved for after work, never on the job.

Such goofing off or horseplay should be avoided on the construction site. Take your work seriously and wait until quitting time to play around.

Tools

Performing high-quality work in any profession requires high-quality tools and proper knowledge of their use. The construction industry is no exception. Workers are often judged by the type and quality of their tools and by their ability to use them correctly. Cheap tools are a sure sign of an amateur. Workers should purchase the best quality tools they can afford. The initial purchase is the "rock" on which they expand their assortment of fine tools and instruments. Properly cared-for tools can last a lifetime.

Screwdrivers, for example, should be chosen based on the type of work for which they will be used. The efficient holding power of a screwdriver depends on the quality of steel in its blade, the design of the blade, and the external force applied to the screwhead. The blade should be fitted to the width of the screw slot for best results.

If a common double-wedge screwdriver (figure 3-2) is used in a deep screw slot, the blade transmits its torque (pressure) to the top of the screw slot. This pressure can damage the screw and even cause a section of the screwhead to break off. Wedge-shaped tips also force the screwdriver out of the screw slot—again damaging the screwhead (figure 3-3).

But a screwdriver tip ground to properly fill a screw slot removes any type of screw without a problem because the torque is applied at the bottom of the screw slot where the screw is strongest. Also, the blade fills the slot completely (figure 3-4).

Quality is important for all tools. Tools meant to cut, for example, should be sharp. Dull tools can cause more accidents than sharp tools. Sharp tools are more efficient and allow workers to do their jobs better.

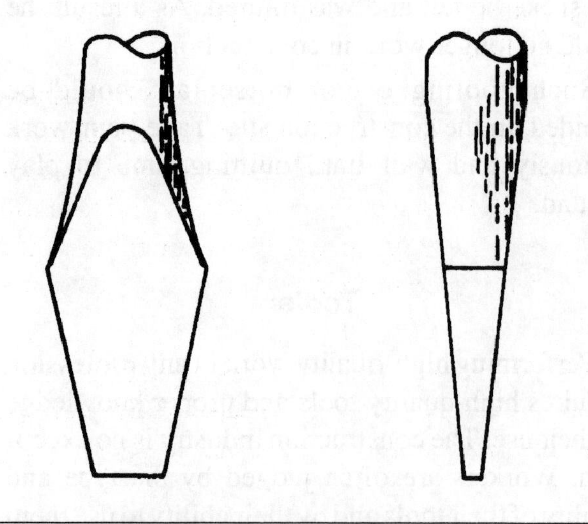

Figure 3-2: The common double-wedge type screwdriver (front and side view) is not the best choice for most construction work.

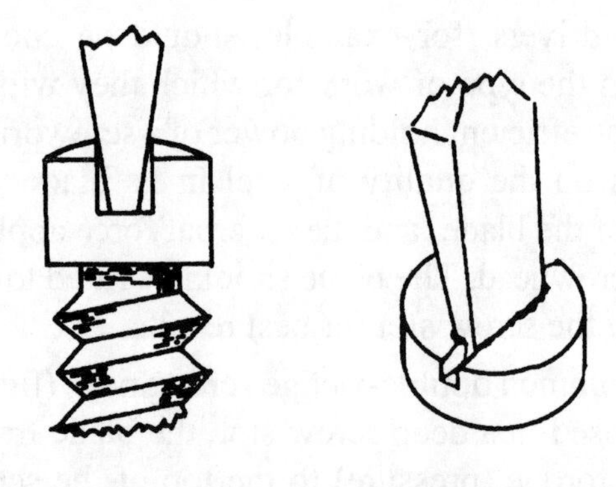

Figure 3-3: (left) Here's the double-wedge screwdriver in use. Note the lack of close-fitting — the blade transmits its torque only to the top of the screw slot. As can be seen, this screw slot (right) was damaged by improper blade fit.

Clothing

The clothing worn by construction workers varies from job to job and depends on the type of work. For example, technicians making final control connections and performing tests in a nearly finished office building with an operating HVAC system will probably wear slacks and a sport shirt;

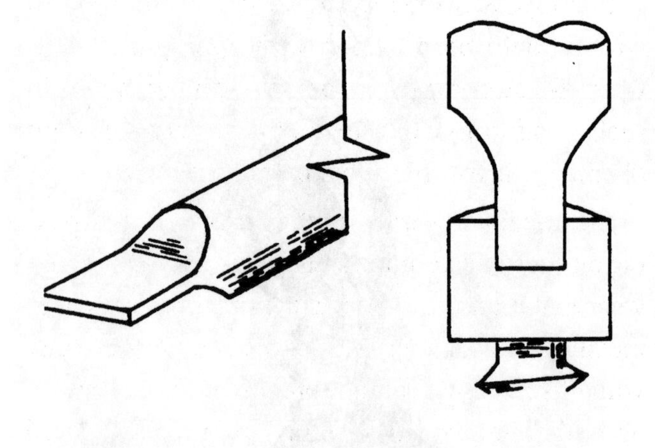

Figure 3-4: This particular blade is ideal. It has been cut to conform perfectly to the slot.

technicians installing the motor controls on the deck of a high-rise building during cold weather will probably wear insulated underwear, sweaters, and coveralls. In all cases, clothing should never be frayed, torn, or otherwise unsuitable for the work and job conditions (figure 3-5).

All shoes worn on the job should have heavy soles and be in good repair. Many manufacturing plants require workers to wear "hard-toe" shoes. These shoes protect toes in case something heavy is accidentally dropped. It is also a good idea to wear shoes containing steel plates in the soles. Workers often step on nails driven into wooden boards. Steel-plated shoes prevent injury from occurring in these accidents. The bottom of the sole should be rubber or another insulated material to prevent electric shocks from energized circuits. For example, electricians frequently replace damaged electrical components on 120- and 240-volt systems. If an electrician is standing on a grounded concrete floor and accidentally brushes against an energized conductor or other component, the electric current will discharge through the body, causing possible injury or at least an uncomfortable shock. Rubber-insulated shoes prohibit the person from absorbing the current provided that no other part of the worker's body is in contact with a grounded object.

Employees working in damp locations or in water should be provided with appropriate footwear. Most contractors provide rubber boots in such conditions. If they don't, workers should obtain a pair for their own personal safety.

Electrical workers and other craftspeople working on energized equipment and circuits should not wear metal articles such as key chains, rings, and metal hard hats. These objects can come in contact with "live" electrical equipment and cause injury or death. The same precautions should be observed around any type of rotary equipment such as motors and fans.

Gloves will protect a worker's hands from many minor injuries such as blisters, cuts, and splinters. However, gloves should never be worn near rotating machinery such as drill motors and fan-coil units, because they can easily get caught in a drill bit or fan blade and severely injure the worker's hand.

Finally, employees who work in tight quarters or are exposed to falling objects should wear an approved fiber or plastic hard hat.

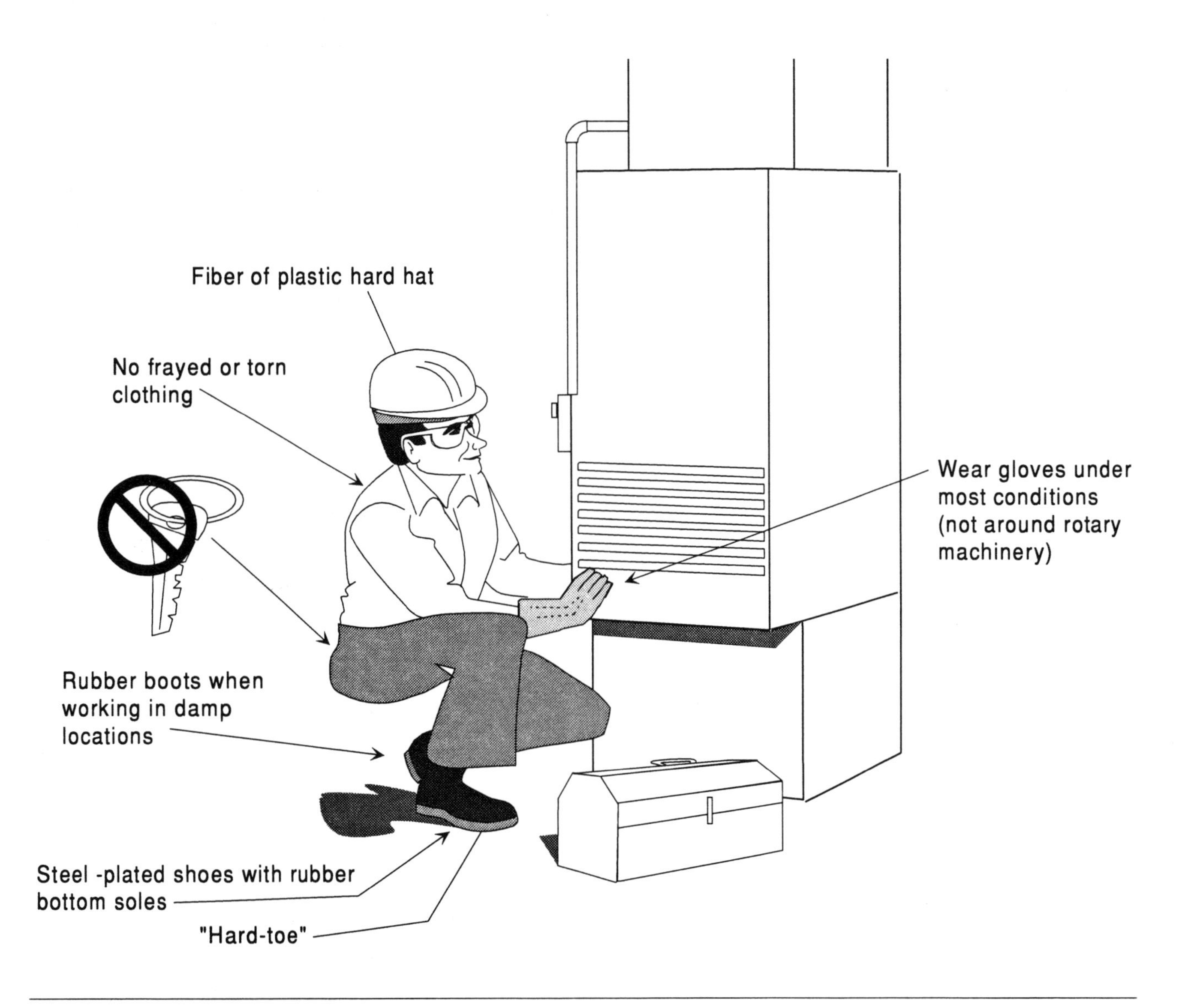

Figure 3-5: Do's and don'ts of clothing for construction workers.

Chapter 4
Construction Safety

Construction sites are hazardous places to work, but a thorough understanding of safe work practices and procedures will help you avoid injury and accidents.

Every worker should be concerned about safety—both on and off the job. Your failure to adopt recommended safety procedures can result in serious injury to yourself and your fellow workers and can cause costly damage to equipment and property.

We often think that accidents happen to others, not to ourselves. But the construction industry, perhaps more than any other, requires constant dedication to safety. Be aware of all potential hazards and stay alert to them. Figures 4-1 through 4-14 are sketches of some DOs and DON'Ts in the construction industry.

Think of safety as a personal attitude. It will help you to work a full career without a serious accident or injury.

SIGNS

Be aware of and understand all warning signs on the construction site, and follow their instructions.

For example, a danger sign—signified by white lettering in a red oval on a black rectangular background, as shown in figure 4-15—indicates areas or machines that pose immediate hazards to workers and equipment. When this sign is encountered, the instructions must be followed exactly in order to avoid injury.

Caution signs have a yellow background with black lettering. The word CAUTION is always printed at the top of the sign with the caution message below. For example, a caution sign stating "Eye Protection Required" may be posted in areas where metal or fiber pieces swirl through the air. (See figure 4-16 for examples of other caution signs.)

Safety First signs are similar to caution signs: They offer warnings and suggestions that will help prevent accidents. This type of sign has an all-white background. SAFETY FIRST is superimposed in white on a green background, with the message below in black letters. Figure 4-17 illustrates some common Safety First signs.

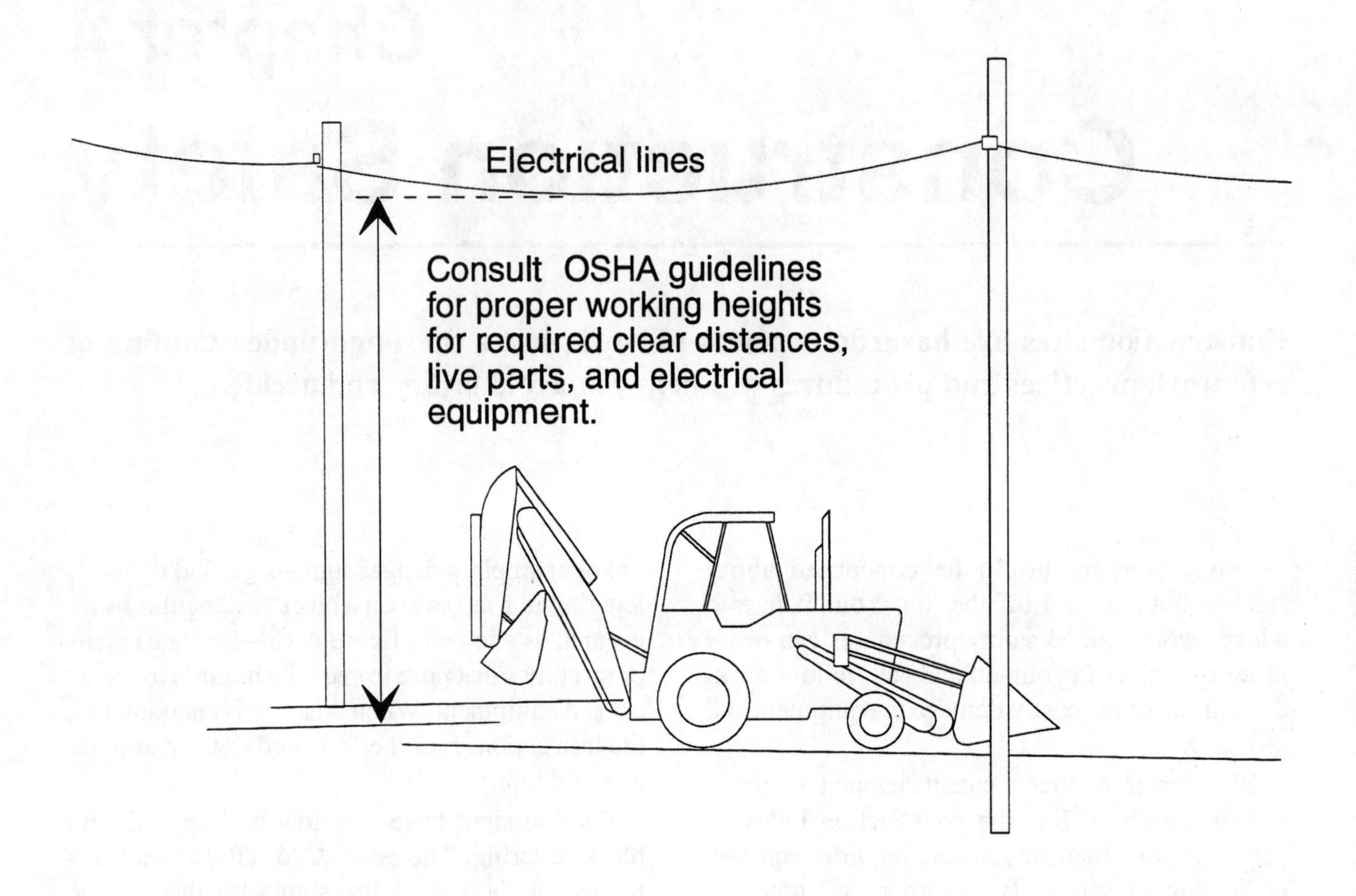

Figure 4-1: When doing any type of digging, watch out for electrical lines—both overhead and below grade.

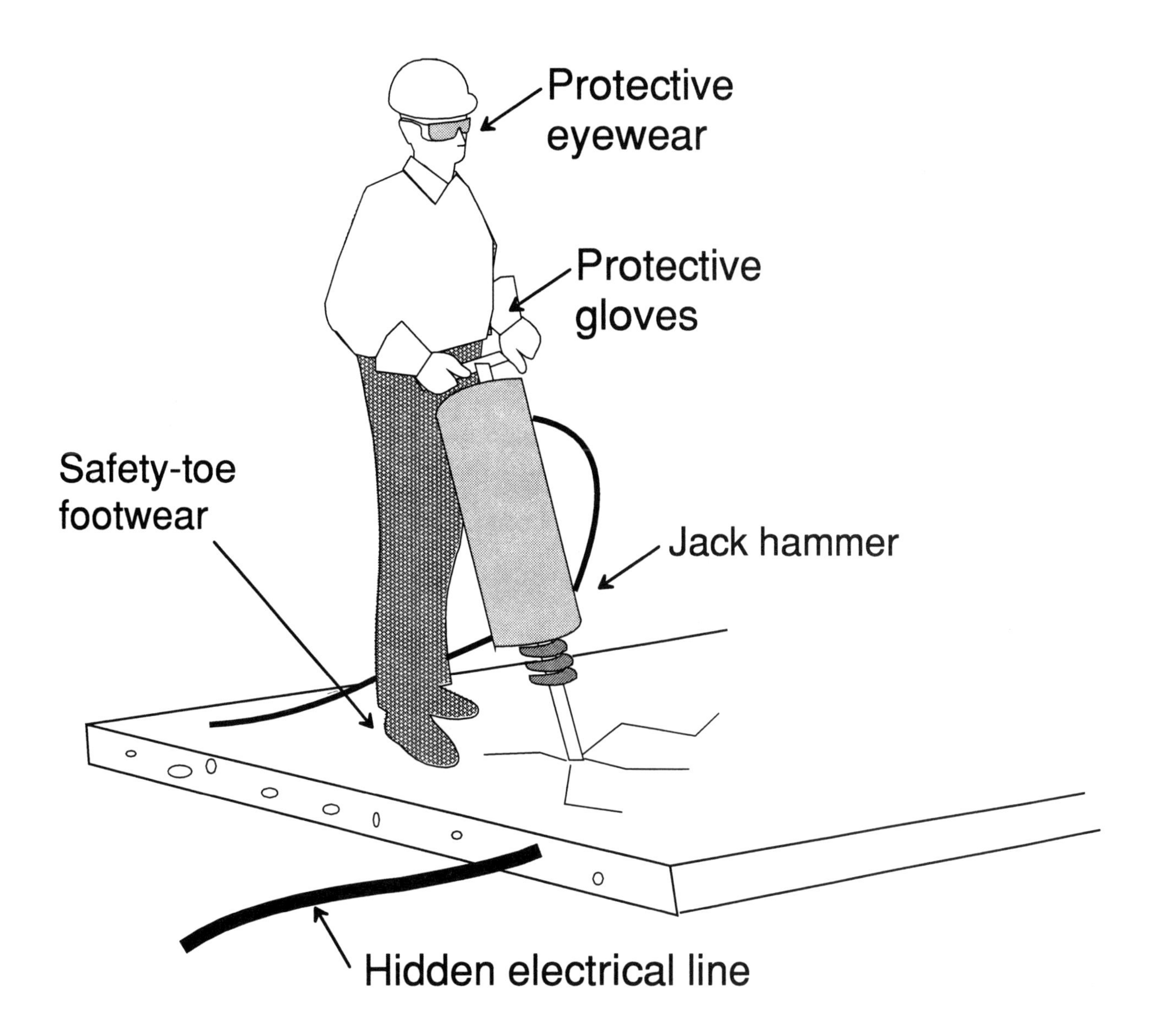

Figure 4-2: Jackhammer operators and others who must dig in existing earth or concrete must be on guard at all times.

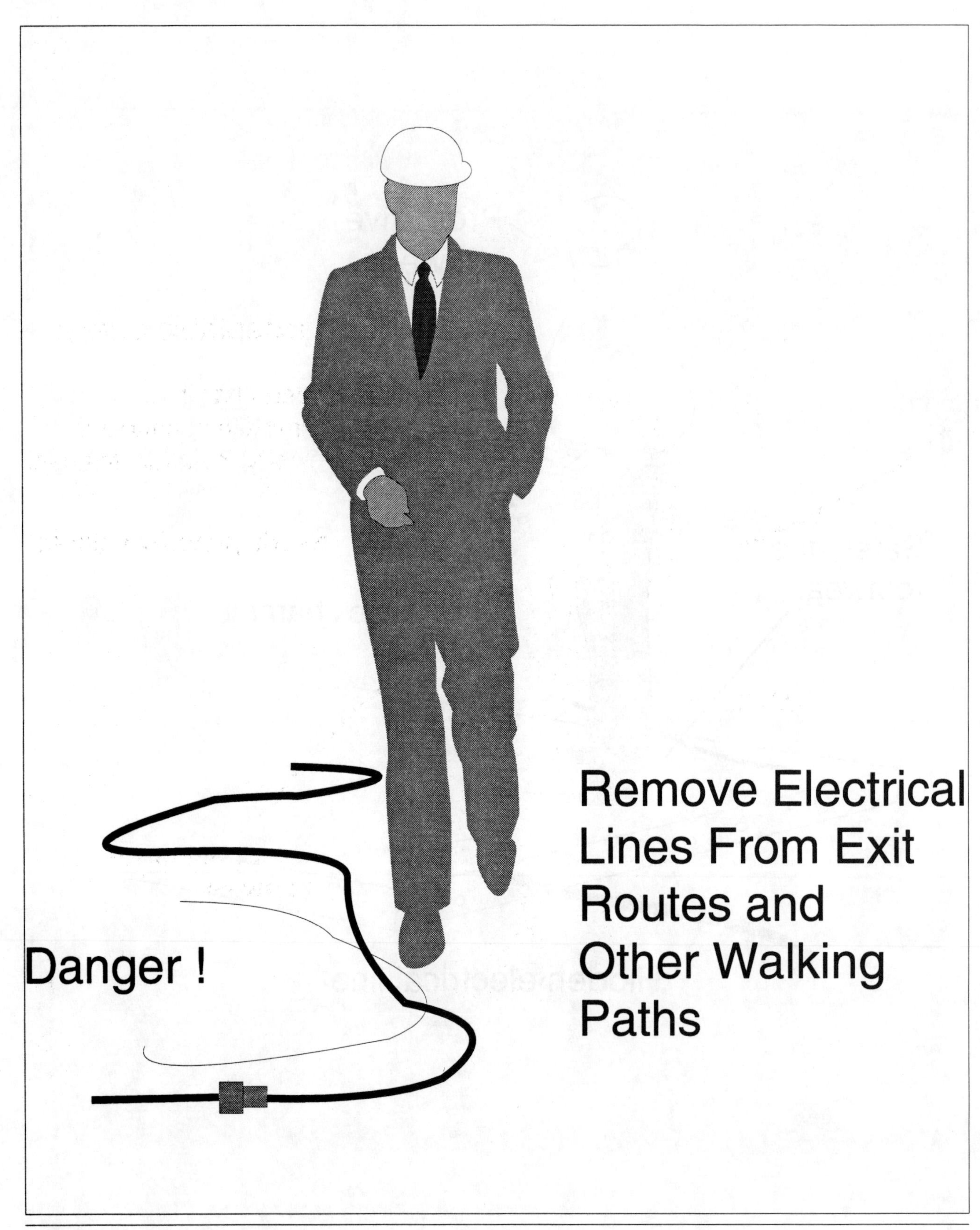

Figure 4-3: Keep unnecessary electrical lines and other debris off the floor. Guards should be installed when such lines are absolutely necessary.

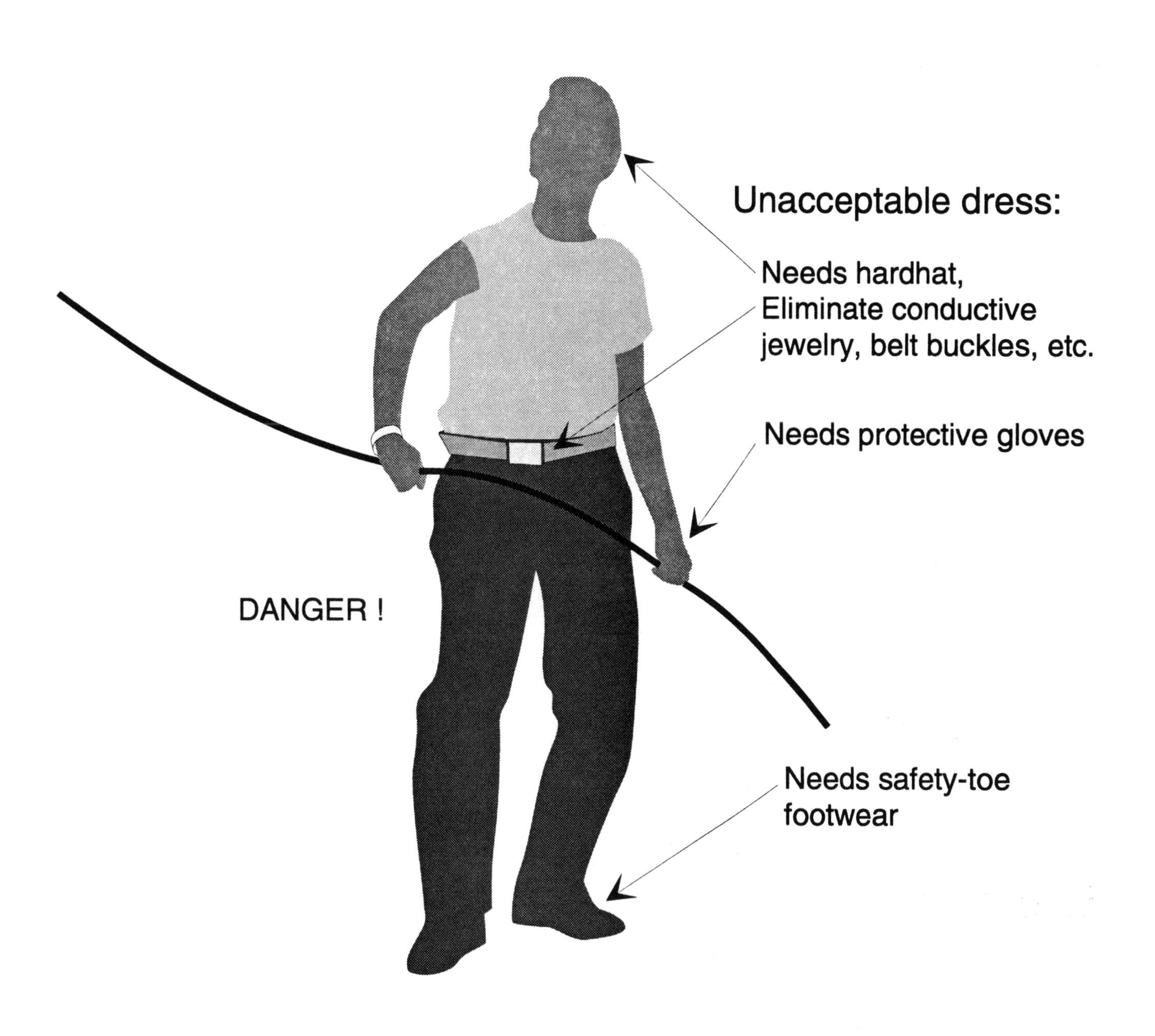

Figure 4-4: Be prepared to handle all construction-related equipment properly.

Figure 4-5: Many workers injure themselves or other workers when carrying ladders, lengths of pipe, and similar items. Know the appropriate methods used to carry all construction tools and materials.

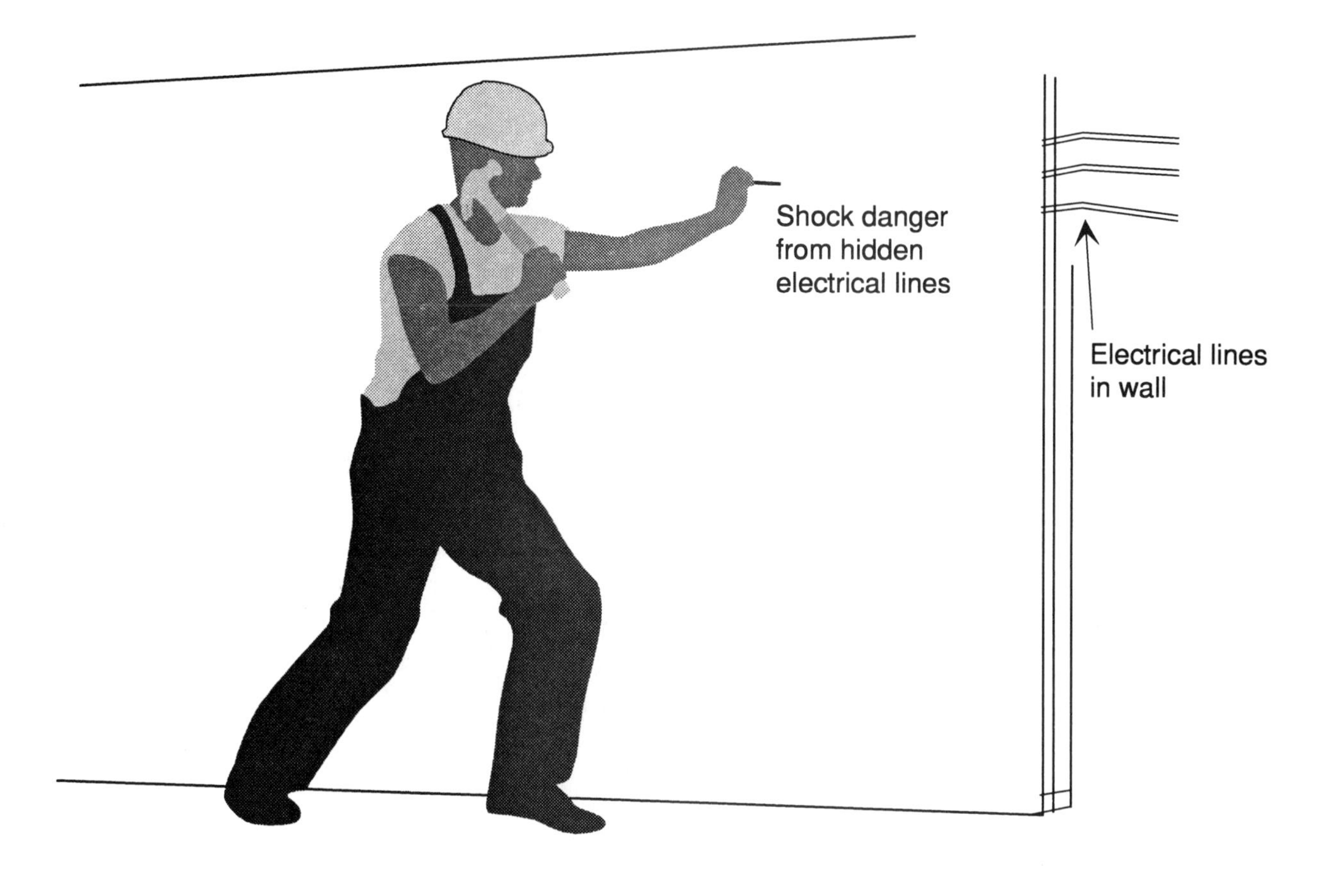

Figure 4-6: All construction workers must be aware of the dangers of hidden electrical lines.

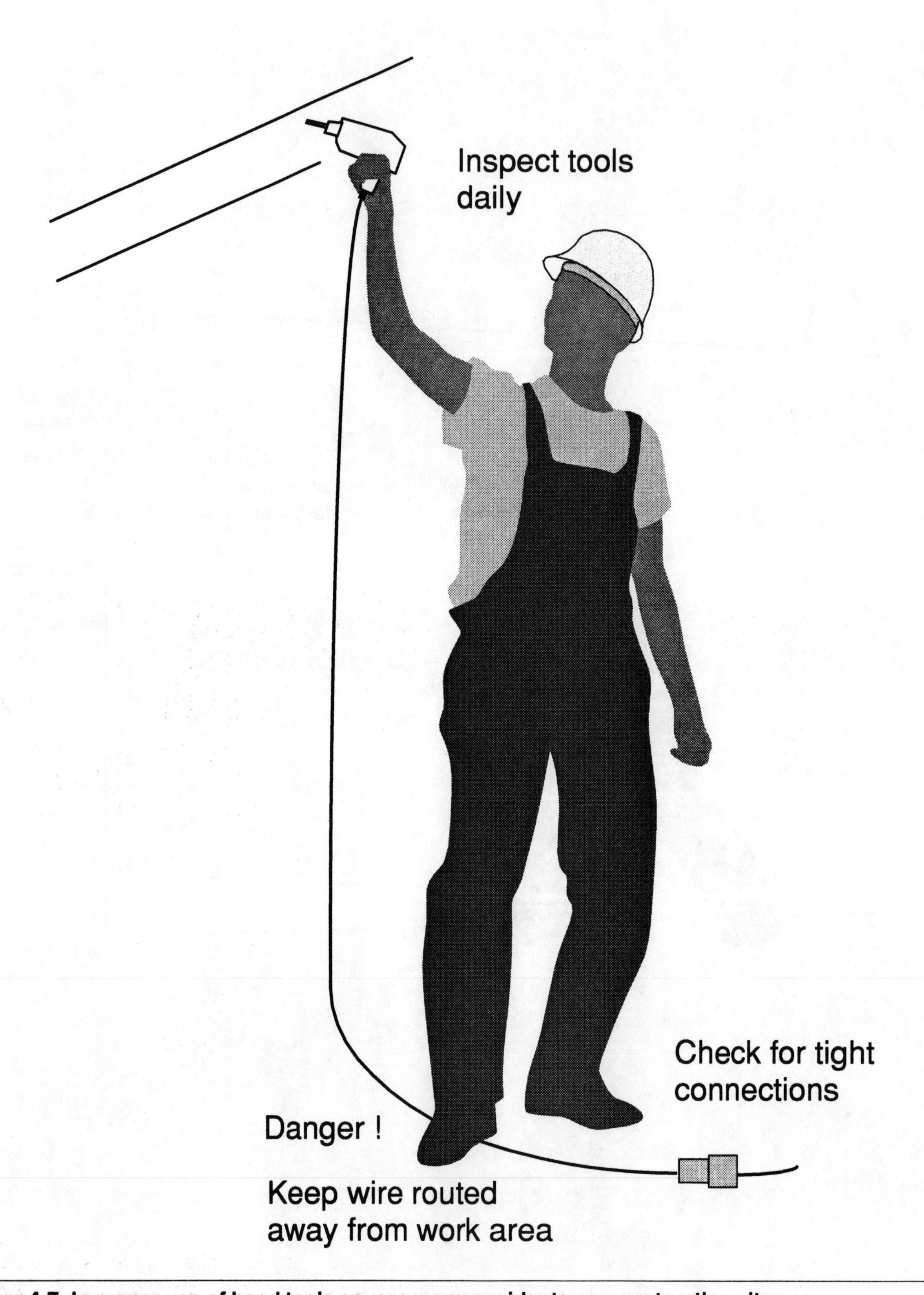

Figure 4-7: Improper use of hand tools causes many accidents on construction sites.

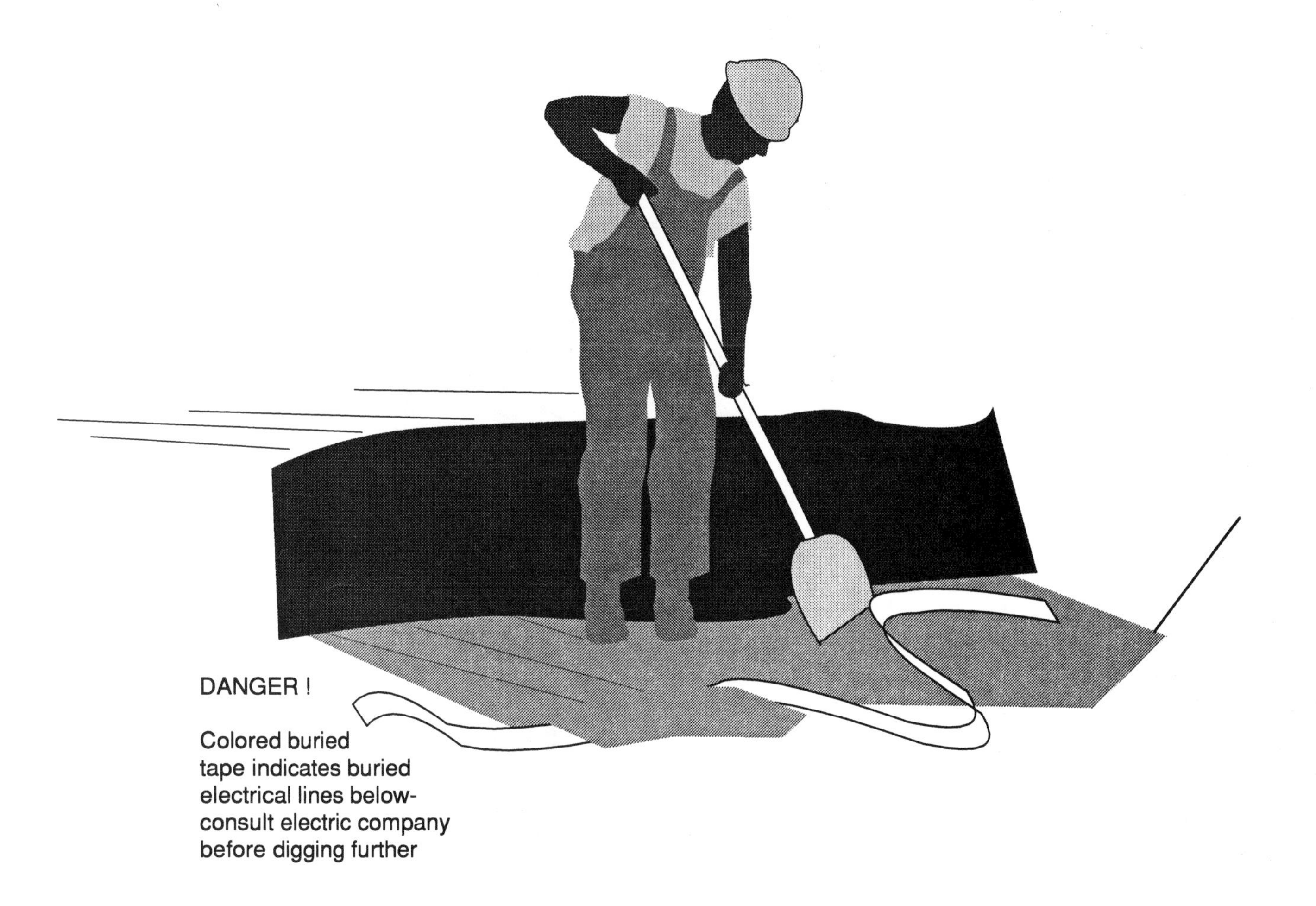

Figure 4-8: When any type of markers, like colored ribbon, are encountered while digging, stop work immediately. "One call" systems exist so all utilities (gas, electric, phone, and so on) can be checked before digging begins.

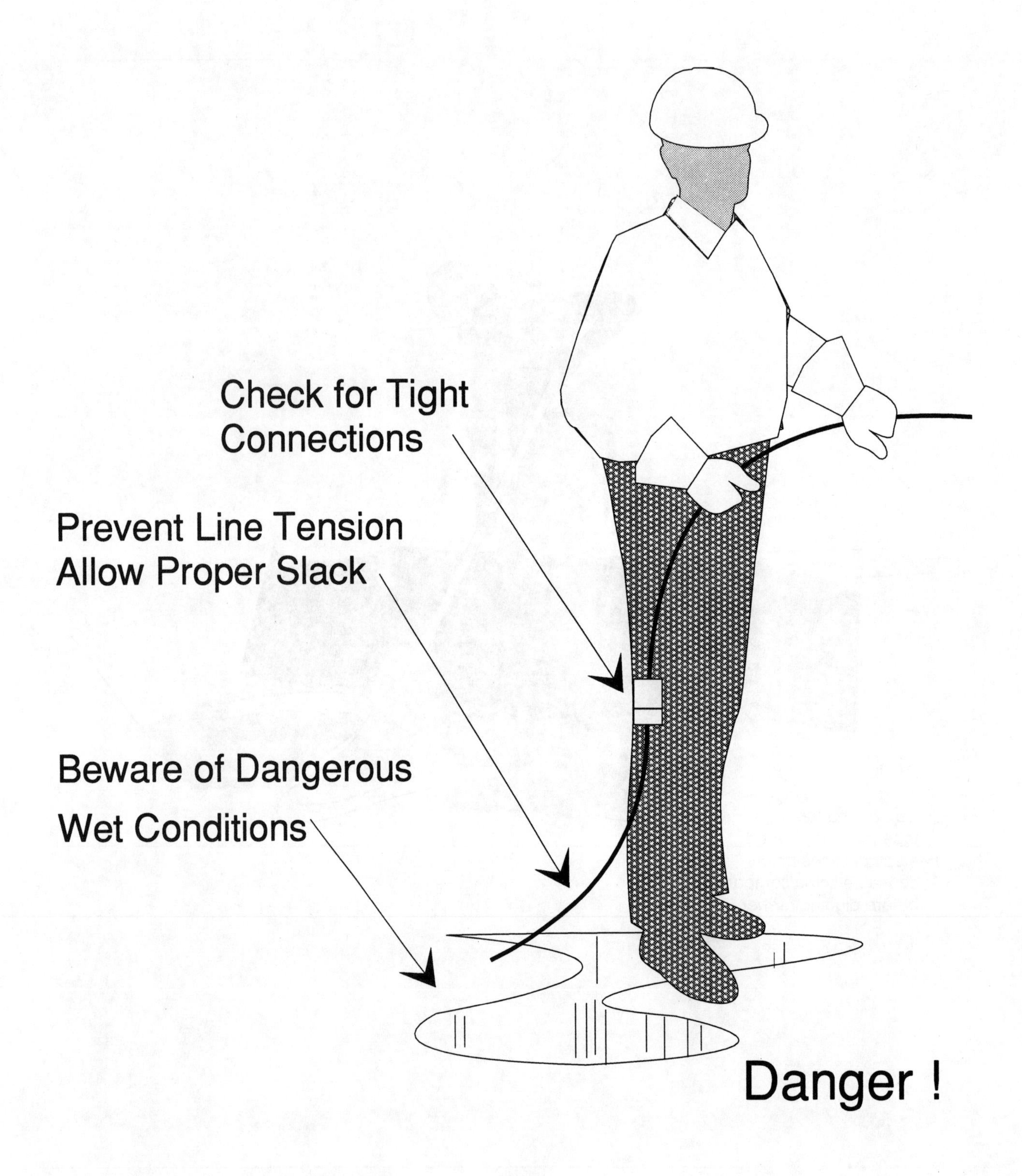

Figure 4-9: Beware of dangerous wet conditions on job sites.

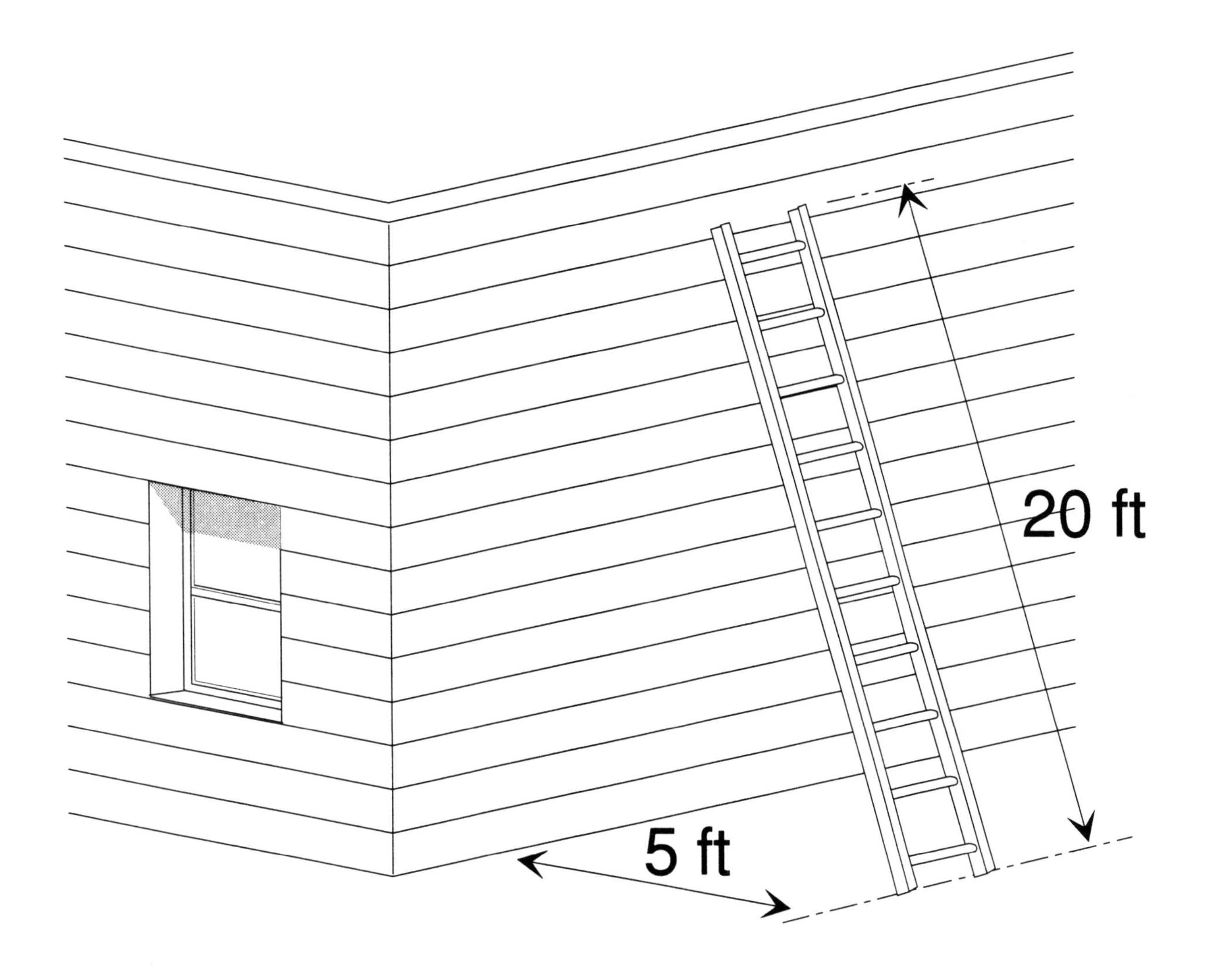

Figure 4-10: When using ladders, the base should be positioned out from the vertical surface not less than one-fourth of the height.

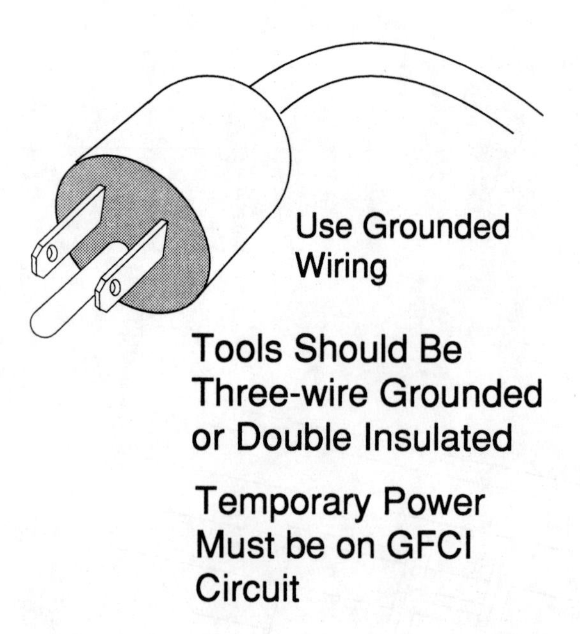

Figure 4-11: Make certain that power cords for portable hand tools have equipment grounding capabilities.

Figure 4-13: The tag-and-lock method can be a life saver on certain construction projects.

Figure 4-12: Equipment should be de-energized and locked-out before performing work on it.

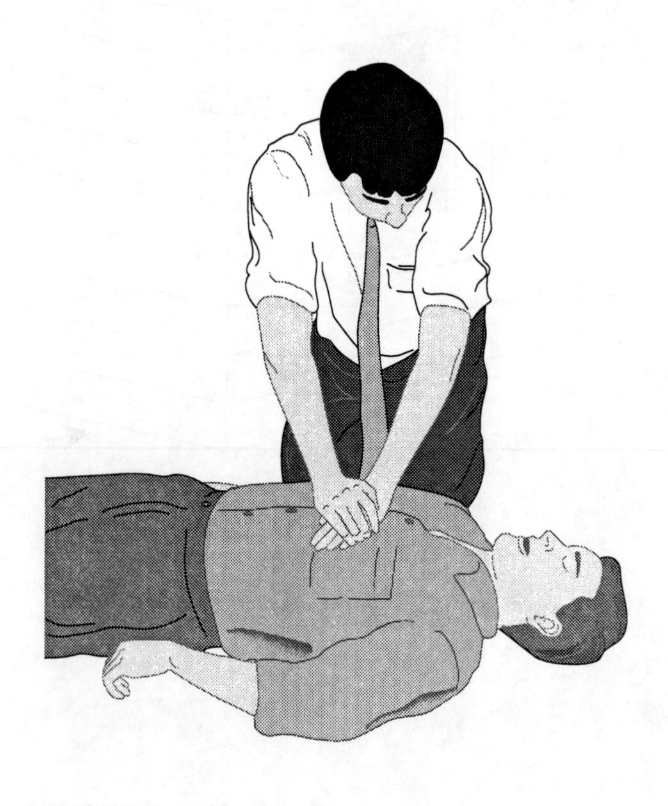

Figure 4-14: At least two members of each work team should be trained in CPR techniques.

Figure 4-15: Some of the many danger signs that will be encountered on construction sites.

CAUTION

CHEMICAL
STORAGE

CAUTION

DO NOT WALK
ON CONVEYORS

CAUTION

EAR
PROTECTION
REQUIRED

CAUTION

EYE
PROTECTION
REQUIRED

CAUTION

FIRE LANE
KEEP CLEAR
AT ALL TIMES

CAUTION

FLAMMABLE
LIQUIDS

Figure 4-16: Sample caution signs found on construction job sites.

SAFETY FIRST
KEEP THIS AREA SAFE AND CLEAN

SAFETY FIRST
KEEP ALL AISLES CLEAR

SAFETY FIRST
GOGGLES REQUIRED

SAFETY FIRST
ALL INJURIES NO MATTER HOW SLIGHT MUST BE REPORTED TO YOUR FOREMAN AT ONCE AND BE TREATED AT THE FIRST AID ROOM

SAFETY FIRST
DO NOT ENTER UNLESS WEARING SAFETY EQUIPMENT

SAFETY FIRST
DON'T TRY TO LIFT MORE THAN YOU ARE ABLE

Figure 4-17: Sample safety first signs found on construction sites.

SCAFFOLDS

Scaffolds are commonly used in construction. They are meant to provide safe, secure, elevated work platforms for personnel and materials. Scaffolds should be designed and built to meet high safety standards. But normal wear and tear or overuse can weaken a scaffold and make it unsafe. It is therefore important to inspect all parts of a scaffold before each use.

Three types of scaffolds are used in the construction industry: manufactured, rolling, and suspended scaffolds.

Manufactured scaffolds are made of painted steel, stainless steel, or aluminum, because these materials are stronger and more fire resistant than wood. They come in ready-made units that resemble the sections of a fence. The individual units are assembled on site.

A rolling scaffold is a manufactured scaffold with wheels on its legs so that it can be moved easily. The scaffold wheels are fitted with brakes to prevent movement while work is in progress.

A suspended scaffold is a platform supported by ropes or cables that are attached either to the top of some support structure or to beams extending from the side of a support structure. The platform is raised or lowered by pulling on the suspension ropes or cables using a hand crank or an electric motor.

Inspecting Scaffolds

A scaffold assembled for use should be tagged with either a green, yellow, or red tag.

A *green tag* identifies a scaffold that is safe for use and meets all Occupational Safety and Health Administration (OSHA) standards.

A *yellow tag* identifies a scaffold that does NOT meet all standards.

A yellow-tagged scaffold may be used, but workers must wear a safety harness and lanyard when they are on it. Other precautions may also apply.

A *red tag* identifies a scaffold that is being erected or taken down. Workers should never use a red-tagged scaffold.

Don't rely on the tags alone. Check the scaffold for bent, broken, or badly rusted tubes. Check for loose joints where the tubes are connected. Such problems present hazards that must be corrected before the scaffold can be used.

Make sure you know the scaffold's weight limit before using it. This weight should be compared with the total weight of the workers, tools, equipment, and material that will be put on the scaffold. Scaffold weight limits must NEVER be exceeded.

If a scaffold is higher than four feet, make sure that it is equipped with top rails, midrails, and toe boards. All connections must be pinned. Cross-bracing must be used. The working area must be completely planked. Cross-bracing does NOT eliminate the need for handrails.

If there is room under the scaffold for people to pass, the space between the toe board and top rail must be screened in to prevent tools and materials from falling off the work platform.

Manufactured or rolling scaffolds that are taller than four times the dimensions of their narrow base should be tied to the structure or guyed (wired) to the ground. For example, if a scaffold with a base of 4 feet by 6 feet is taller than 16 feet, it should be tied or guyed. See Figure 4-18 for safety requirements for scaffolds.

ELECTRIC SHOCK

Electricity is created when electrons move from a voltage source through a conductor that forms a complete path for the movement. This movement is called electric current. Silver, copper, steel, and aluminum are excellent conductors of electricity. Though less efficient, the human body is also a conductor.

Electric current flows back to its source along the easiest path. When the human body becomes

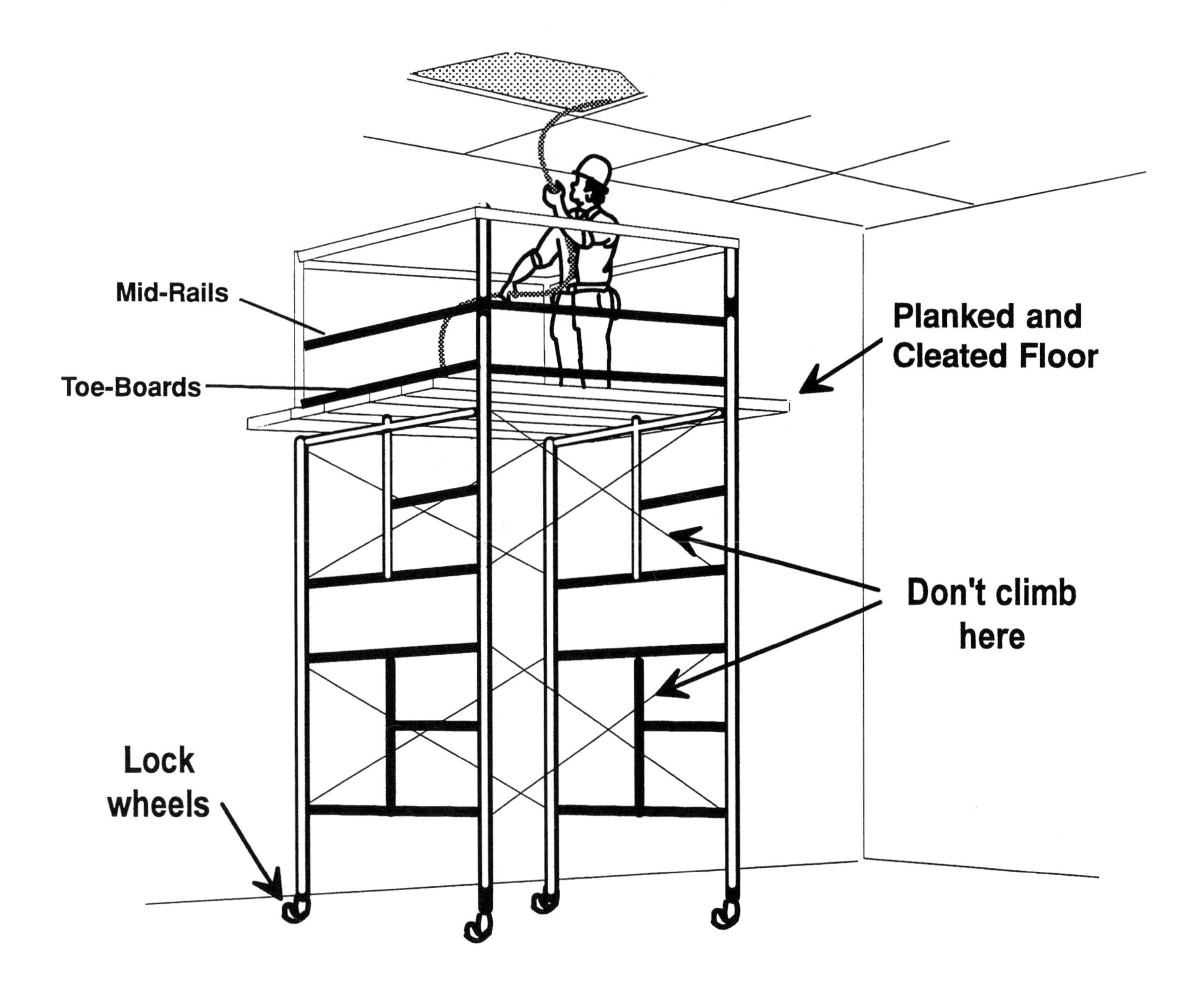

Figure 4-18: Some safety precautions to be used with scaffolds.

the path, it receives an electric shock. If the shock is strong enough, it can be fatal.

Electric current is measured in amperes and milliamperes (a milliampere is one one-thousandth of an ampere).

The human body begins to feel electric shock when current flow approaches 1 milliampere. Between 1 and 5 milliamperes, the body feels a tingling sensation. Above 5 milliamperes, painful shock occurs, and at 50 milliamperes or more, severe shock occurs than can cause death.

Construction workers must be aware of all electrical sources on the job site and must be extremely careful around them. High-voltage sources are more dangerous than low-voltage sources, but either can cause injury.

Body Resistance

The human body has relatively low resistance. For example, the resistance from one hand to the other or from one hand to a foot is about 300 to 600 ohms. However, when the body becomes a part of an electrical circuit, it can conduct a considerable amount of current. How much current depends on the condition of the skin and type of contact that the body makes with the voltage source.

For example, dry skin has an average contact resistance of 350,000 ohms. This relatively high resistance will minimize current flow. The amount of contact resistance, however, also depends on the amount of pressure that the skin makes with the voltage source. The greater the pressure and contact area, the lower the resistance and the greater the possibility of shock.

The moisture content of the skin also affects contact resistance. When the skin becomes wet from perspiration, water, or some other liquid, contact resistance drops from an average of 350,000 ohms to only 1,000 ohms. Again, the actual resistance will be even lower if the pressure and area of body contact increases, and thus the amount of current flow will be considerable. Obviously, working in wet conditions around electrical sources is more dangerous than working in dry conditions.

Very low voltage sources, such as flashlight batteries or even a 12-volt car or boat battery, do not provide enough milliamperes to harm you. This voltage is simply too low to cause any significant amount of current to flow in the human body. However, even 40 volts is enough to cause painful—but not fatal—shock if the body is wet.

Most victims of electric shock are injured by the 120-volt circuit—the most common voltage source. Construction workers sometimes become lax when working around 120-volt lines, but use extreme caution when working with 480-volt circuits, for example. In reality, both voltages can kill.

SAFETY AND YOUR INCOME

Life and health are priceless to everyone. But the simple fact is that if an injury causes you to miss work or become permanently disabled in some way, your income will be affected.

When you lose time, you lose income, but you continue to have expenses. Workers Compensation Insurance rarely pays enough to cover your losses. Your savings can be destroyed, and you can fall behind in your career. Lost time is not easily recovered.

Employability

If you want to change jobs, potential employers will often want to know about your safety record. Unsafe workers are costly to employers, as well as a danger to themselves and other workers.

The information in this chapter is not meant to scare you or discourage you from the construction trades, but rather to help you understand and appreciate the need for safety precautions. These procedures will help keep you safe and protect the property you are working on throughout your career.

Chapter 5

The Building Trades

There are many primary trades in the construction industry, offering a wide variety of challenging opportunities for the future construction worker. All these trades hold the promise of a challenging and lucrative future.

CARPENTERS

Carpenters make up about one-quarter of all skilled craft workers in the construction industry. Carpentry is one of the oldest crafts, and its basic principles have changed little with the passing of time. There will always be a place in the industry for a skilled carpenter.

Carpenters cut, fit, and assemble wood and other materials. They work on residential, commercial, and industrial construction projects that range from building frames for houses (figure 5-1) to building forms for pouring concrete foundations (figure 5-2). Because of their many abilities, carpenters have many opportunities for employment.

Carpenters' responsibilities include the following:

- Installing and finishing trim.
- Constructing and installing kitchen cabinets.
- Constructing fences, decks, and porches.
- Framing and installing walls, partitions, windows, and doors.

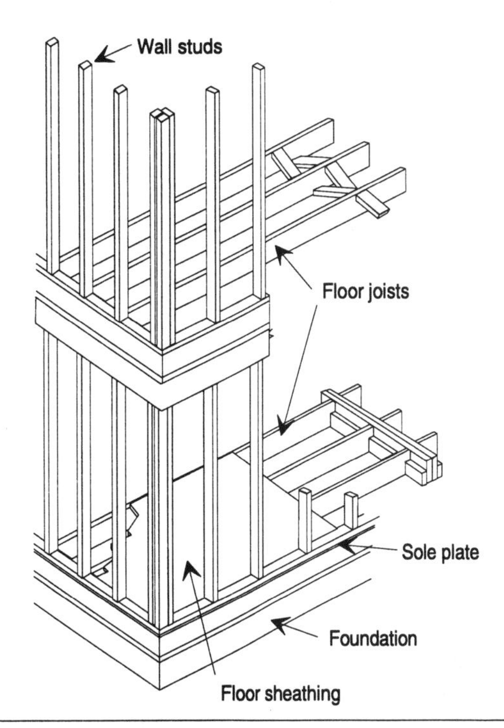

Figure 5-1: Wood-frame construction is handled almost entirely by the carpentry trade.

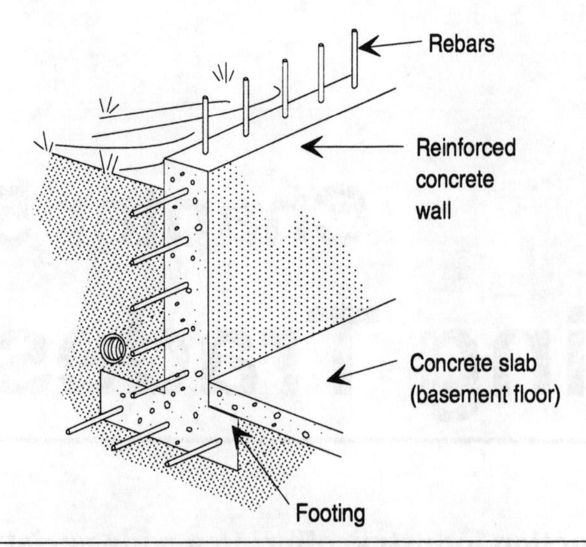

Figure 5-2: Forms for reinforced-concrete construction are usually built by carpenters.

- Providing maintenance and upkeep to existing structures.

Skilled carpenters are masters at using numerous tools, reading blueprints, and changing existing structures (figure 5-3). They must be physically fit and able to communicate with other trades throughout the construction process. They must have a complete understanding of the construction industry.

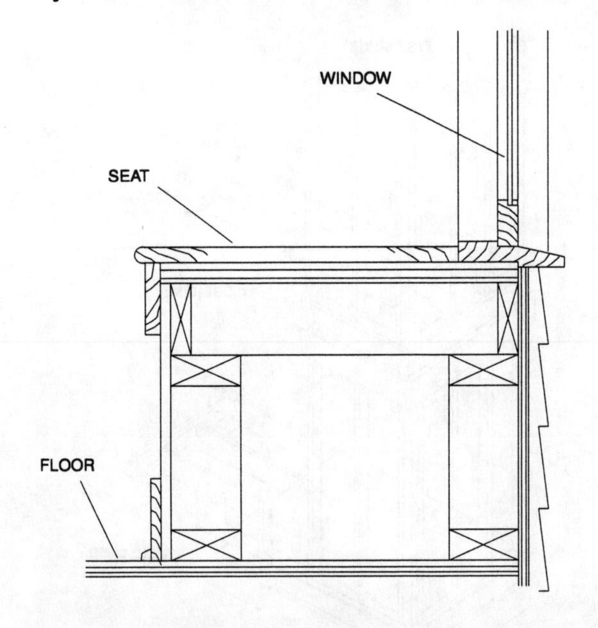

Figure 5-3: Besides the skills required to cut and make accurate joints, carpenters are also required to interpret working drawings and written specifications.

In some cases, carpenters come up through the ranks of the trade by helping experienced carpenters build numerous projects. Many carpenters today have no formal training. However, apprenticeship programs are available for anyone who wants to learn the trade.

Getting into an apprenticeship program is not easy. To qualify, you must be working in the trade and must apply to a state-approved construction organization for admission to a program. These organizations will review your qualifications and decide if you have the ability to master the trade.

Most apprenticeship programs require four years of training, which equals about 8,000 hours of work. Programs include actual work experience in carpentry and 144 credit hours of school training per year.

Some organizations give credit for hours you may have worked before you became an apprentice. Any previous education, such as courses taken at an accredited vocational or technical school, may also count toward the total required credit hours. Carpenters have more opportunities than many other craftspeople to become supervisors.

DRYWALL INSTALLERS

Drywall refers to plasterboard, gypsum board, or wallboard installed in the interior of residential, commercial, and industrial complexes. The drywall installer plans the installation process, installs the framework in which the drywall panels are to be attached, and cuts and installs the panels where required. In most cases, a furring grid is used to support the drywall. The furring grid is a thin strip of wood attached to the joists (support beams) of a wall or ceiling to create an even surface on which to hang the drywall. After the panels have been attached to the framework, a drywall finisher or taper completes the process by sealing and taping the seams between each panel. Once the tape and sealing compound have dried, the drywall finisher sands the surface until it is smooth (figure 5-4). The drywall installer then inspects the drywall to make sure that it has been properly installed and that it has no flaws.

Drywall installers must know where and what type of drywall will be installed, because it is their job to measure the areas in which the drywall will be installed and order materials.

Drywall trade workers must be physically fit because most of the work is strenuous. Panel boards must often be lifted by two or more people, especially in ceiling installations. Panel board is large and bulky; if not handled properly, it will break.

Drywall installers normally start as helpers until they have obtained the necessary skills to work as independent installers. Drywall helpers help installers carry materials to and from the job site and help erect frames and panel boards. There are no

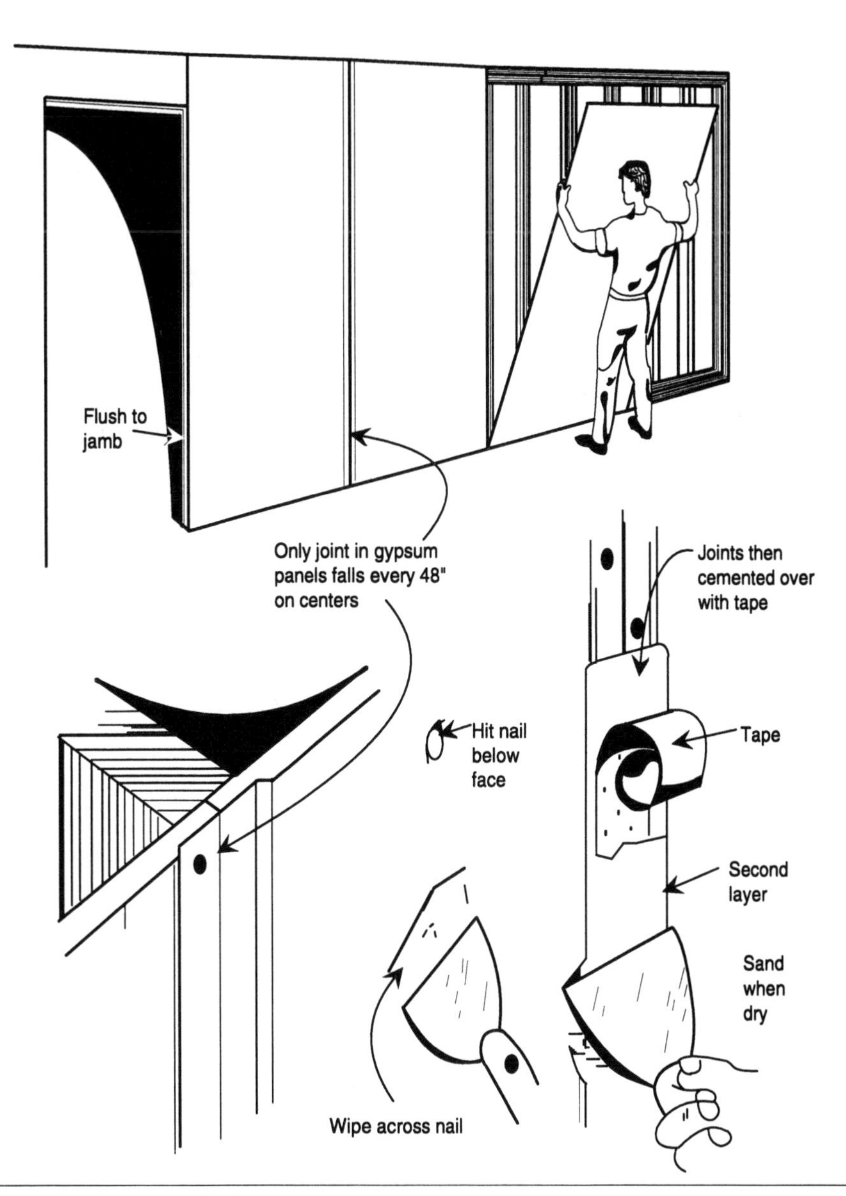

Figure 5-4: Steps required to install drywall.

formal education requirements for drywall installation; however, many companies require these workers to have a high school diploma and the ability to calculate drywall requirements for a given project. There are apprenticeship programs that train workers on the job and provide classroom instruction in such subjects as cost estimating and mathematics.

Helpers and apprentices usually earn about half the salary of experienced drywall installers. Their pay increases as they gain more knowledge of the craft.

Lathers

Lathing is the process by which a structure is constructed to receive plaster, stucco, or concrete material (figure 5-5). The lath covers unfinished spaces of a building, such as the inside framing, with a smooth, solid foundation on which plaster can be applied.

Since the introduction of drywall, the need for lathers has decreased quite rapidly; but because some interior building walls are still plastered, and some older buildings are being restored with their original plastered walls, the trade still exists.

There are seven types of lath materials manufactured today. The most common type used in the construction industry is gypsum lath, which has a plaster base surrounded by a paper-thin surface. Gypsum lath is the easiest type of lath to install.

Many lathers work with construction contractors. Their duties include cutting and installing lathing materials and finishing the inside framework of a building.

Although there are no formal apprenticeship requirements for lathers, the lathing industry recommends at least two years of training. Lathers need some artistic ability, because they may be required to design certain patterns and shapes using specialized tools and materials. These shapes and patterns add to the visual appeal of the lathing process. Classroom subjects for lathers include drafting and geometry. Lathing is not a simple process. Applying lathing materials requires a precise and steady hand.

Plasterers

Plasterers work closely with lathers, applying plaster to the interior walls of buildings after they have been lathed. But like lathers, work opportunities for plasterers are diminishing rapidly because few buildings use plaster anymore. Drywall installers are fast taking the plasterer's place in modern building construction. However, plasterers can command top wages when their skills are needed.

ELECTRICIANS

Electricians work throughout the construction industry, providing electrical services in residential, commercial, and industrial projects. Electricians have one of the highest employment rates of all craft workers. Electricians must be licensed in most areas of the United States and must pass a written exam. All electrical installations must comply with the National Electrical Code® (NEC), and the written exam asks many questions that deal with this code. Copies of the code may be obtained from most electrical inspection offices or directly from the National Fire Protection Association, Batterymarch Park, Quincy, MA 02269.

Electricians install new wiring and related components, such as breaker boxes, switches, light fixtures, and even telephone and television wiring. Some electricians also install motors and air conditioning or heating systems. They perform maintenance and repairs, replacing defective parts as problems occur and repairing equipment when it fails.

Electricians are exposed to many hazardous conditions and situations. However, they can avoid serious accidents or injuries by being aware of potential hazards and staying constantly alert to them.

The electrician must plan and lay out electrical systems around the work of other trades. Often the requirements for electrical installations are not spelled out in the project's plans and written specifications; therefore, the electrician must be able to develop electrical plans to meet the needs of the consumer. When the requirements are included in the specifications, electricians must make sure that the plans are current.

Electricians must be carefully trained. State-approved apprenticeship programs provide a combination of classroom and on-the-job experience in which trainees work with other skilled electricians (see figure 5-6). These programs are by far the best

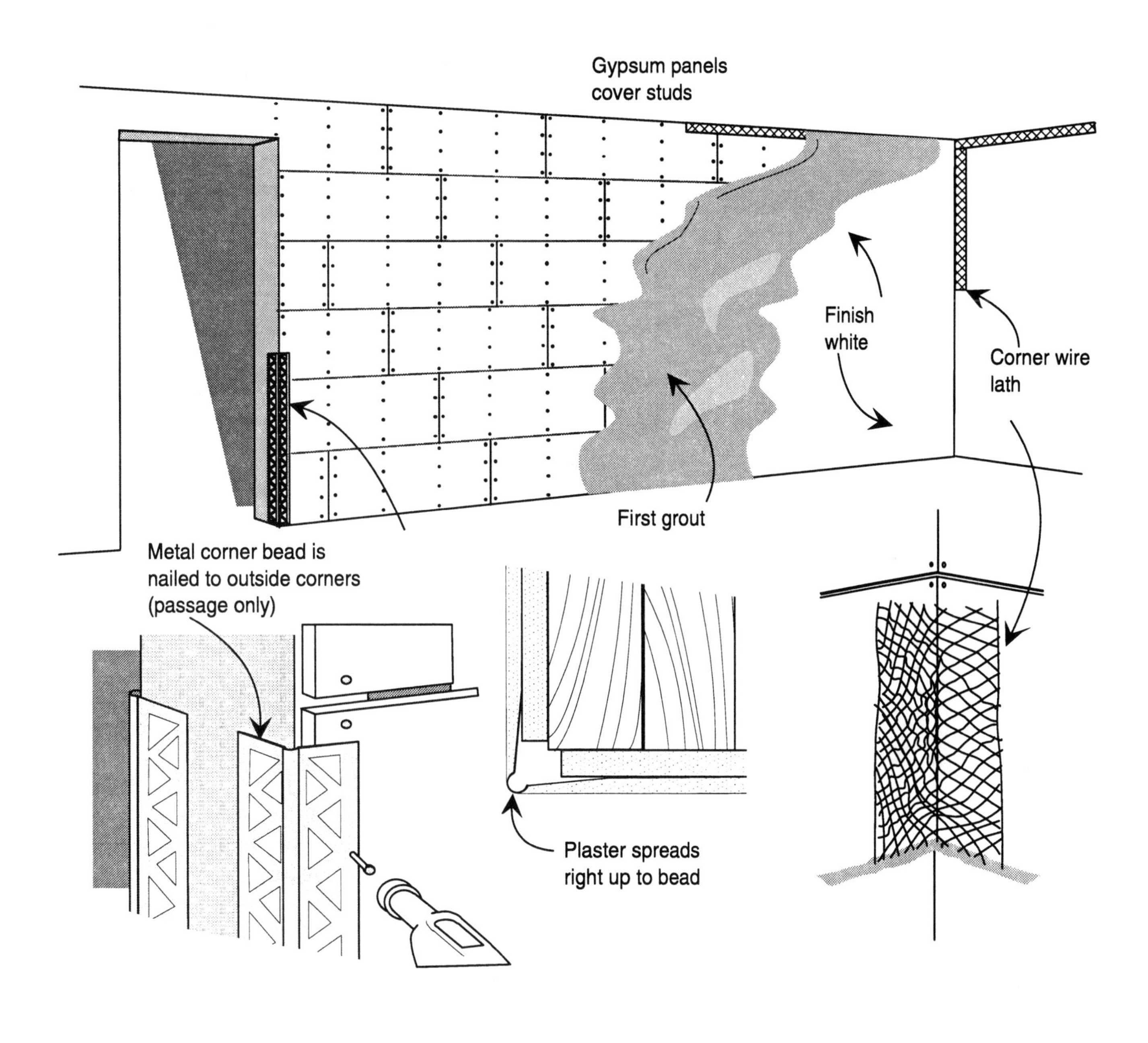

Figure 5-5: Steps required for lathing and plastering a building wall.

- Principles of direct current
- Principles of alternating current
- Basic electrical circuits
- Wire identification and application
- Wire connections
- Grounding
- Overcurrent protection
- Boxes and fittings
- Rules and regulations
- Wiring methods
- Electric services and panel boards
- Electric motors and motor controls
- Tools and supplies

Figure 5-6: Typical electrician apprenticeship program.

To become licensed, electricians are required to take a written examination.

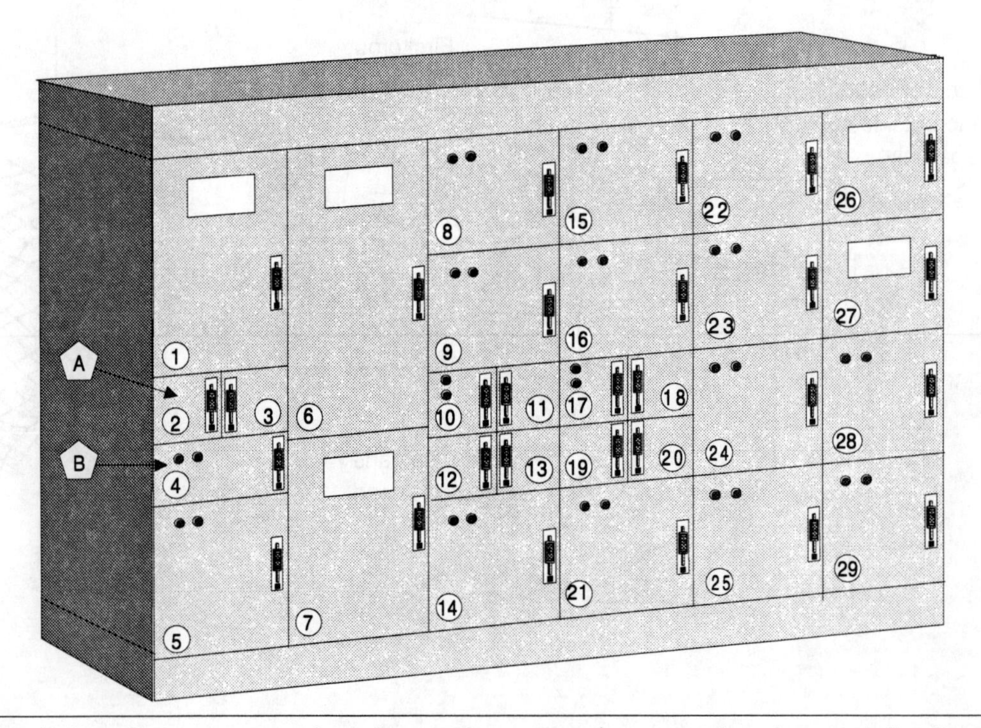

Electricians will install many types of equipment for generating, transforming, and distributing electricity.

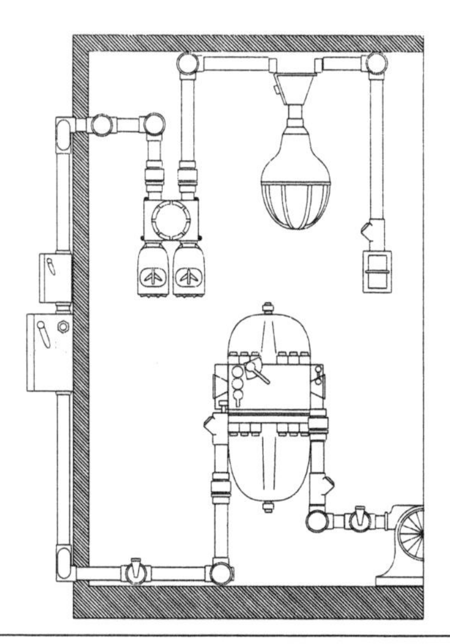

Electricians must have a good knowledge of print reading to perform their work on any electrical installation.

Communications is another interesting field in which the electrician will work.

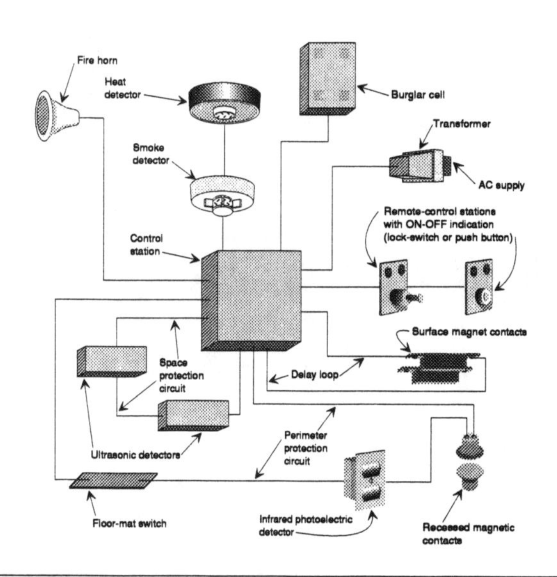

Installing security/fire-alarm systems requires workers with a good knowledge of electrical/electronic techniques.

way to learn the trade and to keep abreast of NEC requirements and changes.

They provide electricians with a foundation that builds on the NEC-approved wiring methods.

Apprentice electricians spend approximately 8,000 hours in training. They are paid about half the salary of journey workers and receive pay increases every six months (journeyworkers are first-class electricians who have completed apprenticeship training).

GLAZIERS

The term glassmaking refers to the production of glass through the process of fusion, in which crystal-like materials are joined by volcanic heat. While still a liquid, the glazed material is formed into the desired shape and size. Glasswork was introduced into the United States in 1608, and the first glasshouse was established in 1739 by Casper Wistar.

In some respects, glassmaking has become a science. Today, there are many forms of glass, including heat-resistant glass, safety glass, and

A large percentage of outside walls for modern high-rise buildings consists of glass — ensuring a good future for glaziers.

fiberglass. These materials are used in many ways and bring beauty to the construction of buildings worldwide. Glaziers cut and install glass as required by the project. The glazier may work in a small shop or at the work site. Regardless of where the job is performed, glaziers must be trained to use the tools and equipment of the trade.

Approved apprenticeship programs for glaziers last about three years or 6,000 hours. In addition to on-the-job training, the apprentice glazier learns the theory of glassmaking and how to handle specialized tools, construct scaffolding, and install numerous types of glass.

Glaziers install glass for windows, doors, and other opening devices (figure 5-7). These trade workers are found in glass production plants and large commercial installations. Very few residential projects require glaziers because most win-

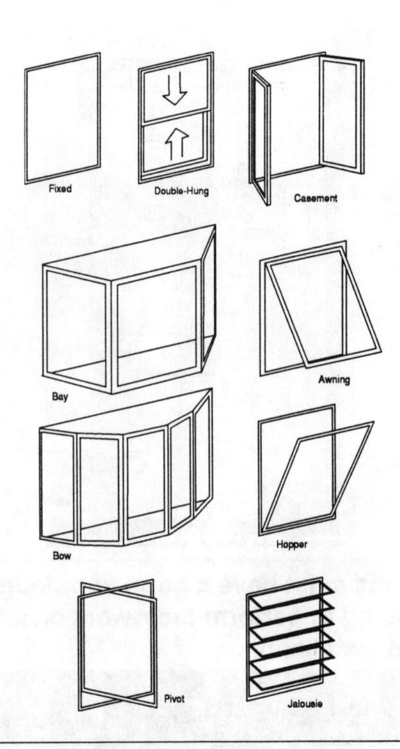

Figure 5-7: Glaziers work with windows and glass of all sizes and shapes.

dows used in residential construction are ready made at the factory. Even renovations of large apartment complexes seldom used customized windows and doors; all are ready-made units. However, in years to come, more residential construction jobs will require the installation of solar glass panels, which may increase the demand for professionally trained glaziers.

HEATING, VENTILATING, AND AIR-CONDITIONING (HVAC) TECHNICIANS

Automatic heating, ventilating, and air-conditioning (HVAC) systems are one of the truly great advances of the 20th century. These systems allow us to set a desired condition on a control or thermostat, and practically forget it. If surrounding conditions drop below the desired temperature, the system calls for just the right amount of heat. If the area is too hot, the system automatically switches

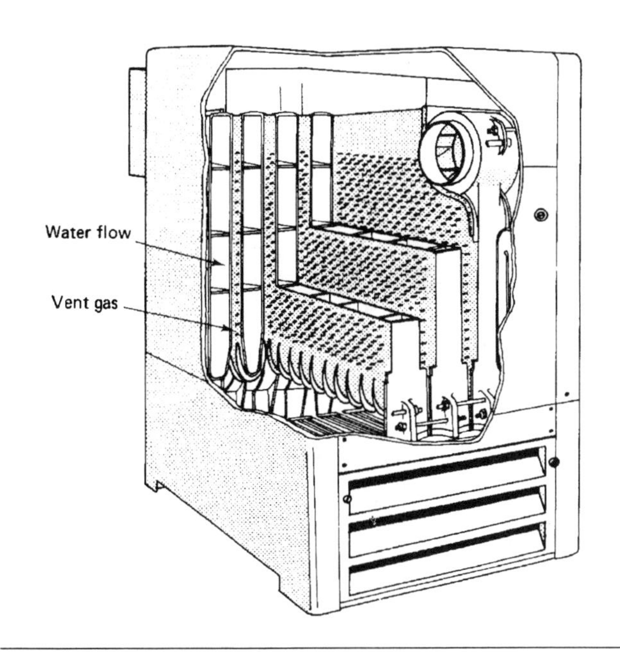

Cut-away view of a gas-fired boiler.

to the cooling cycle. The air can be cleaned electronically and a prescribed amount of fresh air can be introduced into the system.

Although often taken for granted, the fundamental concepts of heating and air conditioning are not understood or even considered by the millions who enjoy their comfort.

Air conditioning changes the condition of the air in an enclosed area. We spend most of our lives in enclosed areas. Thanks to air conditioning, people work harder and more efficiently, play longer, and enjoy their leisure time more comfortably.

The use of HVAC systems in the United States and other nations continues to grow at an unbelievable rate. Most new buildings in the United States have some type of heating and cooling systems; 50 years ago, the luxury of air conditioning was limited to theaters and similar establishments. Many older buildings that did not have HVAC systems have been renovated in recent years and these systems have been installed.

In short, the ability to control temperature has become a priority for society. In addition to pro-

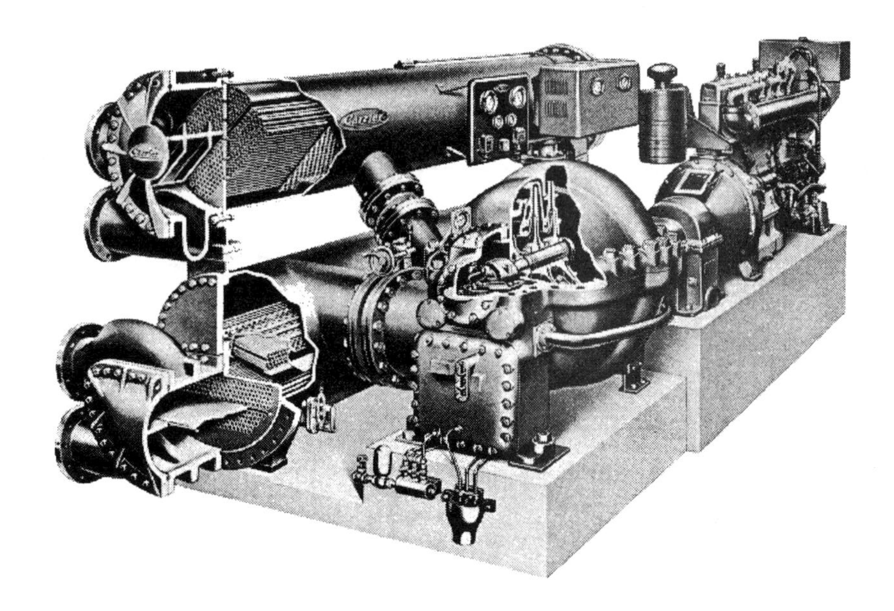

The HVAC worker may be called upon to install projects as small as a residential kitchen exhaust fan to projects such as huge centrifugal refrigeration machines as shown above.

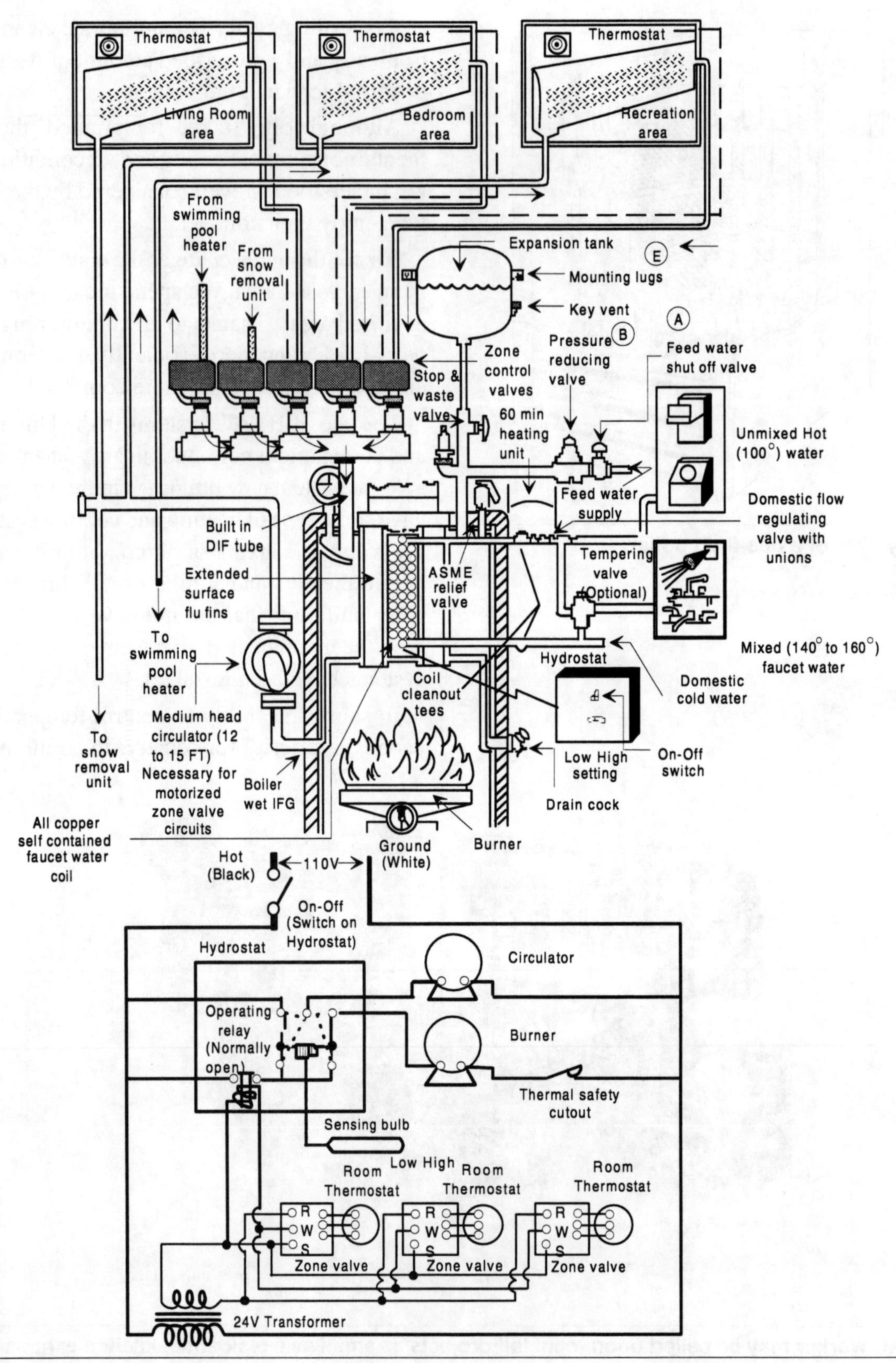

To install HVAC equipment properly, the HVAC worker must have a thorough understanding of the operating principles of the systems being installed.

viding comfort, this ability has permitted such scientific achievements as the following:

1. Military centers that track and intercept hostile missiles are able to operate continuously only because air is maintained at suitable temperatures.
2. Atomic submarines can remain submerged almost indefinitely, partly because of air conditioning.
3. Modern medicines are often prepared in scientifically controlled atmospheres.
4. Human exploration of outer space has been greatly simplified by air conditioning.
5. Refrigeration equipment makes it possible to store food, medicine, and other perishable items safely.

The demand for heating and cooling services has created tremendous growth in job opportunities for HVAC and refrigeration technicians. These trade workers install, maintain, and repair such systems and equipment. HVAC technicians

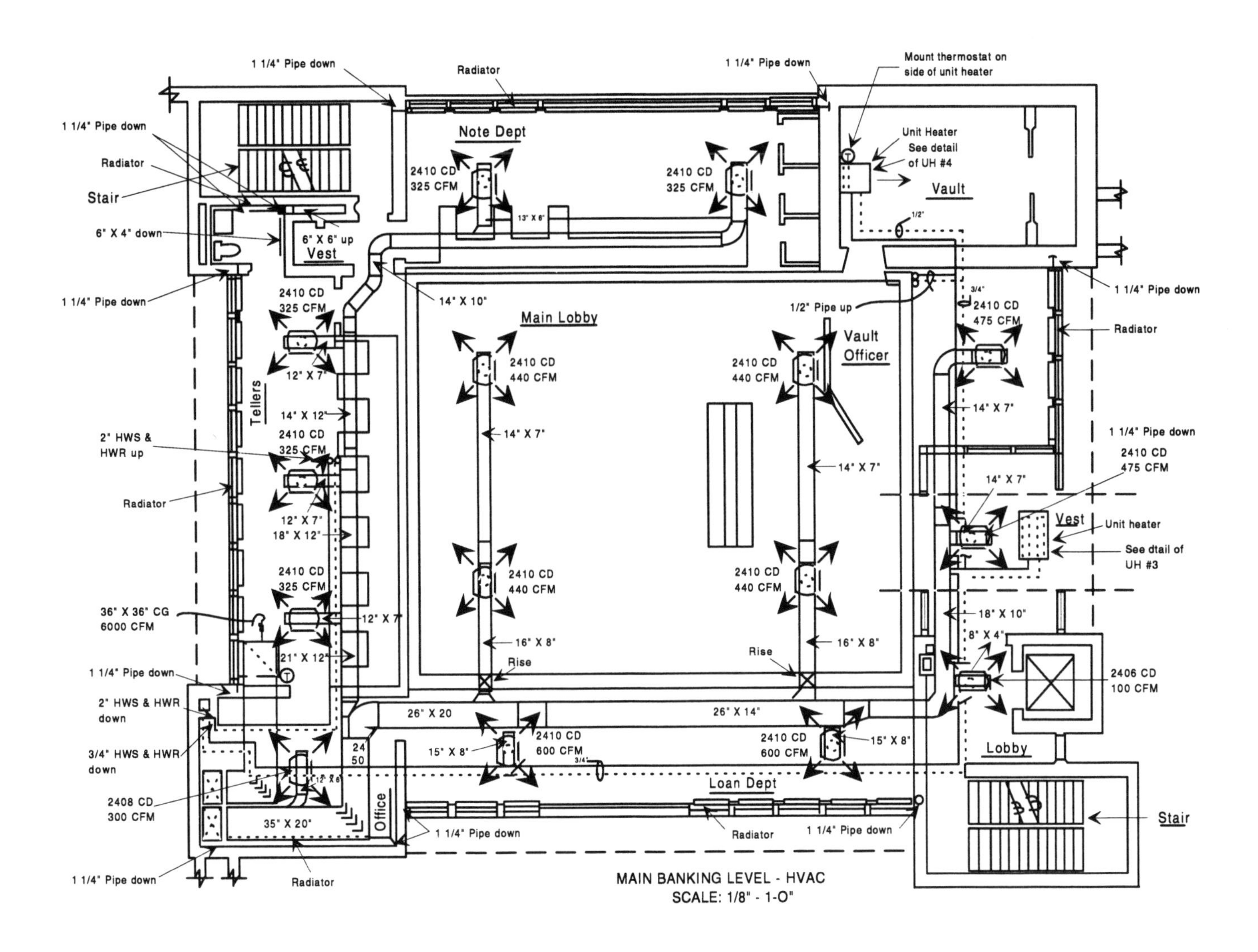

The HVAC worker must have a good understanding of print reading and must also be able to interpret written specifications for all types of projects — from small residential to large industrial establishments.

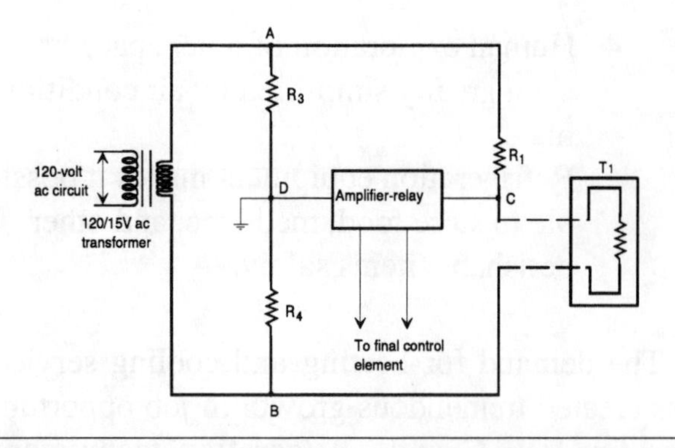

HVAC control work is a specialized field within the HVAC trade. Workers in this field usually command 20 percent more income than the conventional workers.

The HVAC field is not limited to residential and commercial projects. Industrial establishments offer secure employment for qualified HVAC workers.

may specialize in installation or in maintenance and repair. Some work only with certain equipment, such as oil-fired furnaces or commercial refrigeration. However, many cover the entire HVAC and refrigeration field—installing and repairing all types of systems.

Skilled HVAC technicians must have a basic knowledge of HVAC operating principles and must be familiar with the tools and methods used to install HVAC and refrigeration systems.

In the past decade, the demand for HVAC systems has expanded, and the technology of HVAC has advanced. Today's systems are highly energy efficient and thus can save consumers thousands of dollars by cutting their energy costs.

Typical heating and air-conditioning systems include gas, oil, coal, and electric furnaces; electric forced air systems; heat pumps; and hot-water baseboard heat.

Most of the heating systems used today are either gas or oil fired. Technicians who specialize in these systems often have more work than they can handle. They must have stamina and the ability to manage several projects at once. HVAC technicians may own their own business or work for a construction, trade, services, or manufacturing firm; or they may work for the government.

Most HVAC technicians are trained in high school programs and then take additional trade courses. These courses may be three or four years long. Students in high school programs often work as helpers to qualified technicians during weekends and summer breaks.

To keep up with changes in technology and to expand their skills, experienced technicians often take additional courses offered by associations such as the Refrigeration Service Engineers Society, the Petroleum Marketing Education Foundation, Air Conditioning Contractors of America, and Associated Builders and Contractors.

The Bureau of Labor Statistics reports that employment opportunities for HVAC technicians are greatest in residential, commercial, and industrial construction. These opportunities are tied to the construction of new buildings and the replacement of older systems, but the emphasis on energy conservation and the continuing advances in technology put a premium on the skills of HVAC technicians.

Heating and air-conditioning technicians must be able to work with other trades throughout the construction process. For example, duct work for HVAC systems is usually installed by sheet metal workers. As a result, many HVAC contractors hire their own team of sheet metal workers or train their technicians to install duct work and HVAC.

HVAC Apprenticeships

Air-conditioning and heating systems control the temperature, humidity, and even the cleanliness of the air in homes, offices, factories, and schools. Refrigeration equipment allows us to safely store food, drugs, and other perishable items. Air-conditioning, heating, and refrigeration technicians install, maintain, and repair such systems and equipment.

Apprenticeship programs for HVAC technicians sometimes emphasize installation or maintenance and repair for certain equipment, such as gas furnaces or commercial refrigerators. But many provide training in all types of service for all types of equipment.

Skilled technicians learn the basic skills of both hand and machine operations and the advantages and limitations of each. Then they learn to adapt this knowledge to specific projects. When technicians start a job, they use their knowledge and skills to solve any problems they encounter.

Technicians can become supervisors for construction firms or on various projects. Some open their own contracting businesses; however, it is becoming increasingly difficult for one-person operations to succeed.

Employment opportunities for air-conditioning, refrigeration, and heating technicians are expected to increase in coming years. Many openings will occur as experienced workers transfer to other fields or retire. Even during periods of slow job growth, workers will be needed to service existing systems. Installations of new energy-saving systems also will create new employment. The production and marketing of food and other perishables means that more refrigeration equipment will be needed—and more technicians to service it. At the same time, large industrial plants, which are desperate to conserve energy, will require the services of HVAC technicians to install the latest equipment or keep their current systems running efficiently. These situations will also add to the opportunities available in HVAC.

INDUSTRIAL COATERS

The harsh environments of many industrial establishments destroy certain materials—especially metal—if they are not properly treated. Many specialized paints, fiberglass wrappings, and other coatings have been developed to protect industrial equipment and related systems that are exposed to these environments. The highly specialized industrial coater analyzes, designs, and installs such coatings.

INSULATION WORKERS

The insulation trade is, in some respects, related to industrial coatings.

The term *insulation* refers to a group of natural or synthetic fibers that are fire resistant. These materials are used to insulate piping, duct work, hot water heaters, walls, ceilings, and floors.

Insulation is used to conserve energy by preventing the hot or cool air produced by an HVAC system from escaping. Electric bills can be cut by 25 percent if insulation is used correctly. When

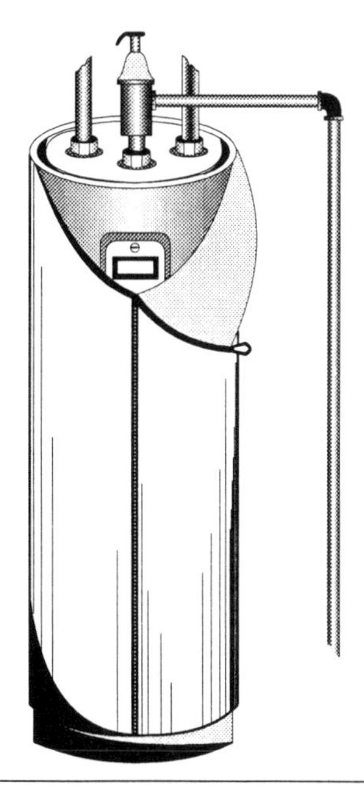

A water heater is one item that is frequently insulated.

nsulation is put around a hot water heater, for example, the heater will not have to work as hard, because the water won't cool as fast.

Insulators typically install insulation in large industrial complexes. Most large petroleum refineries and chemical plants need a lot of insulation around piping that holds and carries hazardous materials. This insulation prevents the pipes from freezing and bursting during the winter.

Sometimes insulators must remove old insulation, which can require a lot of research and careful planning with plant managers and others, especially if asbestos is involved.

Insulators often learn their trade by completing a qualified program. These programs are normally four years long and give a lot of practical hands-on training. The trainee learns to use specific tools during all phases of installation, to estimate the time it will take to complete a project, and to determine the cost of a job. They also learn job management and safety practices. After completing the four-year program, trainees must pass a final examination in order to become qualified journeyworkers.

Many experienced insulators become foremen, supervisors, or estimators for large corporations. These positions require a lot of experience working with all types of insulation in a variety of settings.

INSTRUMENTATION WORKERS

The instrumentation trade involves developing, installing, and repairing instrument devices that are used throughout the construction industry. Workers in this trade specialize as designers, repairmen, and technicians. Each specialty has its own function. The instrumentation designer, for example, repairs or modifies electronic and mechanical instruments, such as timing devices. Using precision measuring instruments, hand tools, and welding equipment, the designer modifies instruments to meet the requirements of construction documents.

Most instrumentation workers specialize in only one area. For example, the parts mechanic works on lighting, heating, and power instruments. These workers enjoy good working conditions, although they must be patient and have a good understanding of mechanical functions. Much work is done using highly sophisticated machines and measuring and testing equipment. Instrument repairmen repair gas or water meters and other devices for utility companies. Many repairmen specialize as meter repairmen, scale balancers, or gas regulators. Gas regulators are in high demand in large industrial plants where meters measure and control such activities as the flow of gas through pipelines.

Most industrial plants and other large construction companies require that instrumentation workers have a high school diploma and at least two years of specialized training through a formal apprenticeship program. Individuals need a strong background in physics, math, and mechanical drawing. Some instrumentation workers start as general laborers or helpers in an industrial plant. Others begin by working with experienced instrumentation workers.

Many instrumentation workers become supervisors overseeing the maintenance and design of instrumentation devices. Others become servicing agents or customer representatives for large corporations.

IRONWORKERS

Bridges, skyscrapers, elevators, and any number of other structures could not be constructed without trained iron- or steelworkers. These tradespeople generally work as a team, and most of their jobs are accomplished outside.

Trained iron- or steelworkers can specialize in structural, reinforcing, or ornamental projects.

Structural ironworkers raise and pull larger girders and beams into place. Projects they work on include erecting bridges, framing, and supports for buildings and equipment.

The working conditions within the iron- and steel-working industry are physically demanding and require the handling of heavy objects and the ability to solve problems.

Reinforcing ironworkers strengthen concrete in walls, piers, roads, and the like. These workers must know when and how to support the concrete evenly.

Ornamental ironworkers specialize in building elevators, stairways, and balconies. They handle heavy preconstructed objects.

Ironworkers must be able to weld, and they must be physically fit. Because they so often work on a crew, they must also be able to communicate with other workers.

Ironworkers are typically trained in apprenticeship programs that take about three years to complete. These programs include on-the-job and classroom instruction and cover such subjects as drafting, mathematics, and blueprint reading. The apprentice learns welding and riveting, as well as how to work as a member of a crew.

Some ironworkers receive their training by working as helpers or laborers for already-qualified ironworkers. Many apprenticeship programs will give credit to trainers who have worked with

All of the reinforcing iron in reinforced-concrete buildings is installed by ironworkers.

Much of the work on bridge construction is handled by the ironworker trade.

qualified iron- and steelworkers. Experienced iron- and steelworkers often move up to crew supervisors.

MASONS

Masonry covers a wide range of craft workers including bricklayers, stone masons, concrete finishers, and plasterers. Some of the more common trades are described here.

Bricklayers

Bricklayers in the construction industry are masters at precision. Using various tools, bricklayers, or brick masons, construct walls, partitions, chimneys, basement walls, and similar parts of a building. In wood-frame structures, bricklayers usually construct concrete block walls that rest on the footings. These walls establish a sturdy foundation upon which the remaining structure will be built. Bricklayers work with brick, block, and panels.

They begin their work after the structure's footings have been poured. They must also order the materials they need. Today, preconstructed wall partitions are available and often used, but bricklayers still do the majority of their work by hand.

Accredited apprenticeship programs for bricklayers run from three to four years, and most major construction companies require this training. The training is both classroom and hands on. Bricklayers must be able to read blueprints and build a structure according to their blueprints' specifications. They must also have good basic math skills. Most states do not require bricklayers to be licensed.

Most bricklayers eventually become supervisors for major construction contracting firms. Many work as estimators determining the approximate cost of specific construction jobs. Some become independent contractors.

The work of bricklayers is labor intensive and requires dedication. Many experienced bricklayers entered the trade as apprentices or helpers to another bricklayer or a supervisor.

Concrete Finishers

Concrete finishers usually finish (level) poured concrete decks, sidewalks, walkways, roadways, and similar projects. They may also build concrete forms. These trade workers earn about the same as brick and stone masons.

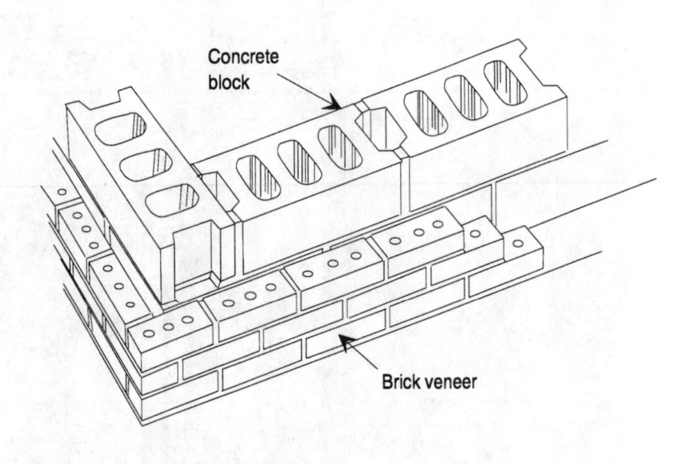

Masonry construction is used to some extent in all types of building projects.

Stone Masons

Stone masonry is much more expensive than conventional brick and concrete block masonry and is used today only in special situations or for decorative purposes. Consequently, work opportunities for stone masons have diminished over the past few decades.

Block Masons

Block masonry is similar to bricklaying. On most projects, bricklayers lay both brick and concrete block. The pay for both bricklayers and block masons is comparable.

Apprenticeship programs for masons usually cover various masonry trades.

METAL BUILDING ASSEMBLERS

Metal buildings began finding their way into the construction industry shortly after World War II. The first metal buildings were quickly and cheaply erected, and a facade was sometimes built on the front to hide the metal behind.

Eventually, firms began designing and manufacturing preconstruction metal buildings that were more pleasing to look at. These buildings could be constructed in less than half the time of a conventional building, and often at less than half the cost. They thus became extremely popular overnight for certain farm, industrial, and warehouse uses. Even many commercial establishments used this type of structure.

A large project requires many masons to complete brick, block, and concrete work.

When an owner or contractor orders a metal building, the manufacturer furnishes foundation plans for the type and size of building ordered. The owner or contractor then builds a foundation according to these plans. The manufacturer delivers the framing and siding materials for the building, and metal building assemblers put the structure together in a few days. Once the building is up, the mechanical and electrical systems are installed in the same manner as in nonmetal buildings.

Early metal buildings were erected by carpenters, ironworkers, roofers, and sheet metal workers. An experienced manufacturer's representative usually stayed on the job site until the building was erected to give instructions as needed.

However, as more metal buildings replaced conventional buildings, metal building assembly developed into a trade of its own. Training courses in the trade are now offered by both metal building manufacturers and various construction organizations.

The length and requirements of this training are similar to training for iron- and sheet metal workers. Many metal building assemblers go on to become experts in the design and layout of metal buildings. Others become supervisors of assembly crews for large construction companies.

MILLWRIGHTS

Millwrights install, transport, operate, and maintain heavy machinery, such as rigging devices, hoists, and jacks. Millwrights are also usually skilled in several other trades in the construction industry, which is important to small construction firms.

Large industrial companies often hire millwrights as equipment installers, movers, and mechanics. These companies use heavy equipment daily, so they need to have millwrights on staff. Small construction firms may only use heavy equipment once a week, so they do not have as great a need for millwrights. Instead, a smaller firm is more likely to hire a worker who has the skill of both a millwright and another craft—a bricklayer or carpenter, for example.

Employment opportunities are steady, for the millwright working in large industrial plants. Millwrights working for small construction companies may experience periods of unemployment.

Millwrights learn their trade either by working on the job with a qualified millwright or through an apprenticeship program.

Apprenticeship programs run about four years and include both classroom and on-the-job training. A high school diploma is often required for admission to these programs. Many high schools offer classes in mechanical drawing, electricity, and machine shop operations—all subjects that provide a good beginning for the student who is thinking about becoming a millwright.

Millwrights work in physically demanding conditions. They must be able to handle heavy objects and solve problems. While all trades require some overtime, millwrights—along with electricians—usually put in more overtime than others.

PAINTERS

Painters paint the interior and exterior of any surface. In the early 19th century, painters were at a premium because most paints had to be mixed by hand and applied with only a brush. Today, paints are matched and mixed by machines, and painters can paint faster and with much more precision. Tools, such as the spray gun, have cut the time of painting in half.

Today, paints can be mixed at paint or building-supply stores. This advance provides consumers with a wide variety of color choices and frees the painter from the worry of making a correct color match. Painters must determine what type of paint to apply to a surface. They must consider the impacts of the environment such as the sun and rain, and must choose a paint that can withstand the elements.

Painters are required to use a wide variety of hand and power tools.

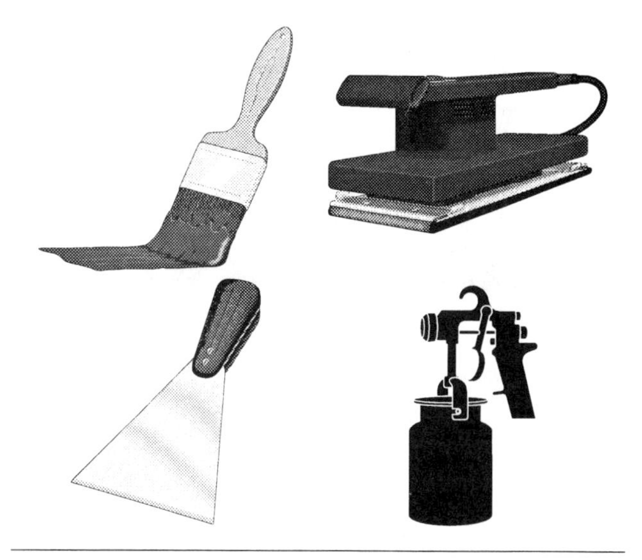

A few of the many tools used by the painter.

Painters use a variety of tools on a day-to-day basis (see figure 5-8).

These tools allow painters to perform their job in an orderly manner. Care must be taken with such tools as the blow torch and chisel.

There are apprenticeship programs for beginning painters, which last about three years. But many painters learn their trade by working for experienced painters. Competition within the painting industry is strong, and aluminum siding improved paints, and other materials have taken their toll. For example, with proper maintenance and touch-up, exterior paint on residential homes can last from five to seven years, so painters cannot hope to stay busy with repainting jobs.

- Putty knife
- Drop cloth
- Electric sander
- Blow torch
- Caulking gun
- Paint mixer
- Wood scraper
- Chisel

Figure 5-8: Tools of the painting trade

Painters who work in dangerous situations, such as on bridges and high-rise office buildings, can expect double the salary of other painters.

PIPEFITTERS

Pipefitters design and install various types of piping systems—hot water, sprinkling, lubricating, and heating or cooling systems, to name a few. They may install new systems or modify older ones. In the past, plumbers installed the piping systems for most residential and commercial establishments. But today, the systems installed in most industrial and commercial complexes require the skills of professionally trained pipefitters.

Pipefitters work with various piping materials and tools and must know the specific use of each. For example, oil, gas, and other chemicals require certain types of pipes and different types of seals and gaskets. Different pipes require different tools for assembly. Pipefitters also need to know local ordinances, state codes, and regulations regarding their work.

Pipefitters typically are trained in apprenticeship programs that run for five years and combine

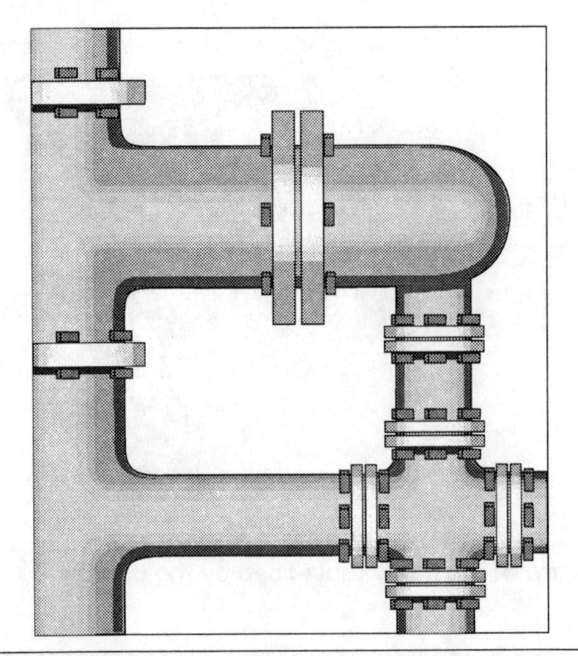

Pipefitters design and install various types of piping systems.

on-the-job training with classroom studies. Applicants must have completed high school with credits in mathematics, general shop, blueprint reading, chemistry, and physics.

Experienced pipefitters can become supervisors for large pipefitting contractors and related companies. Some may decide to become independent contractors.

PLUMBERS

Plumbers design and install piping systems that distribute water and remove waste from a building—for example, washers, bathtubs, sinks, toilets, and heating and cooling systems (see figure 5-9). Plumbers must know about all facets of water distribution, including how to determine the source of water flow and its anticipated water-pressure level.

The National Plumbing Code defines water-supply system as follows:

> The water-supply system of a building or premises consists of the water-service pipe; the water-distributing pipes; and the necessary connecting pipes, fittings, control valves, and all appurtenances in or adjacent to the building or premises.

The water-supply pipe is usually made of galvanized steel, copper, brass, or plastic (PVC). This pipe, which is often called the "main," runs from the water meter, well, or other water source that supplies the building.

The water-distributing pipes branch from the main to serve all fixtures that require water—such as sinks, showers, baths, and drinking fountains. These pipes are usually made of galvanized steel, copper, or PVC, depending on local codes and ordinances.

A plumbing system must usually supply a certain amount of hot water at a given temperature within a specified period of time. The water-heating system may consist of one or more water heaters, either centrally located or placed at various points throughout the building.

Water-heating systems are either one temperature or two temperature.

A one-temperature system provides hot water at a single temperature for all of the building's needs. A two-temperature system provides hot water at two temperatures and is used in certain commercial facilities. For example, restaurants frequently use a dishwasher sanitizing system that requires a very high water temperature. However, this same temperature would be unsafe for the water used in sinks. Two temperatures can be produced by systems that use one or more gas-fired, oil-fired, or electric water heaters, plus the proper mixing valves and controls. The water is heated to the highest required temperature, then this same hot water is mixed with cold water to lower the temperature for the other required level.

Drainage and Vent Systems

Drainage piping removes liquids and sewage from the building. These pipes are usually made of cast iron or PVC and consist of the following parts:

- *Traps*—devices that provide a water seal to prevent odors and harmful sewer gas from escaping through the plumbing fixtures.
- *Vent stacks*—vertical pipes connected to the drainage system that permit a continuous flow of air to the drain pipe. This free circulation of air slows the growth of harmful bacteria in the drainage system and eliminates gases and odors.
- *Cleanout*—a branch pipe from the drainage pipe containing a threaded plug that, when removed, permits access to the drainage pipe for cleaning.
- *Check valves*—devices installed in the drainage system to prevent any liquid from flowing back from the street sewer into the building drainage system.

The number of job opportunities for plumbers in the construction industry ranks second only to electricians. Most plumbers are trained in four- or

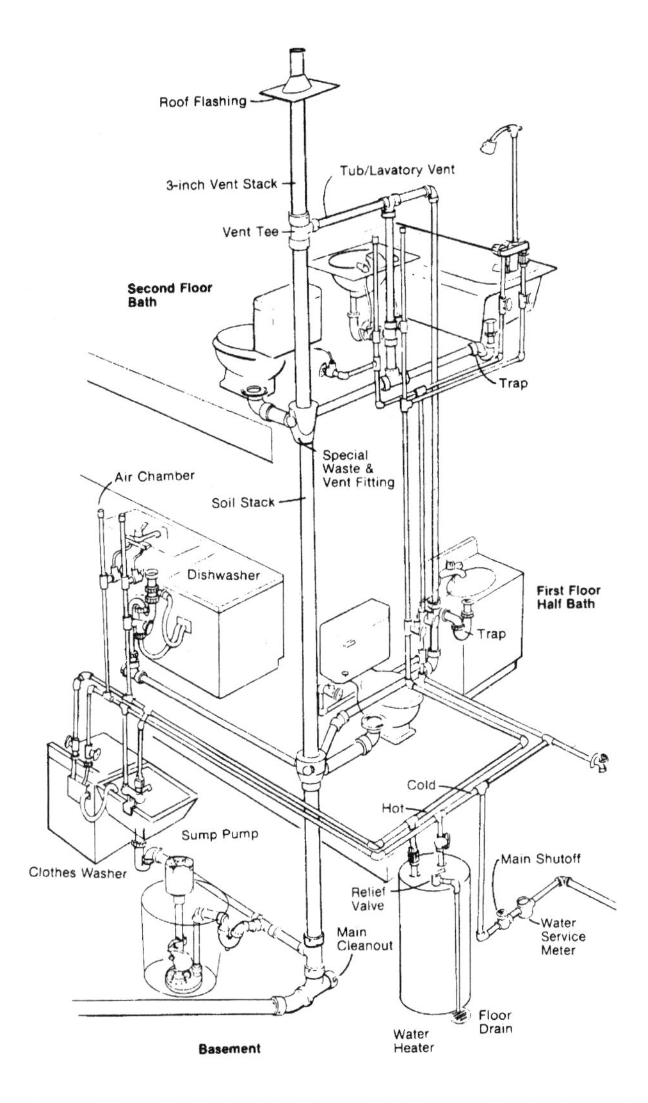

Figure 5-9: The plumbing trade is mainly responsible for piping systems and fixtures that provide incoming water and dispose of liquid waste in buildings.

five-year apprenticeship programs, working alongside a qualified journeyworker plumber. Apprentices receive classroom instruction in blueprint reading, local ordinances and regulations, mathematics, mechanical drawing, physics, and welding and soldering. Many qualified plumbers continue their training by taking correspondence courses and additional classes at trade schools in order to stay on top of changes in the trade.

Plumbers must be licensed to work in most states and must pass a written exam to receive their license. The two-day test lasts about eight hours each day. Most questions are taken from the National Plumbing Code.

Like electrical systems for light and power, the codes and regulations regarding plumbing systems are very strict, because incorrectly installed plumbing systems can cause sickness and death.

SHEET METAL WORKERS

Sheet metal workers most commonly install and repair ventilation and air ducts for heating and air-conditioning companies (see figure 5-10). They spend much time cutting and molding sheets of metal into workable products, and they must be able to select the proper materials needed for their work. Sheet metal comes in many sizes and types.

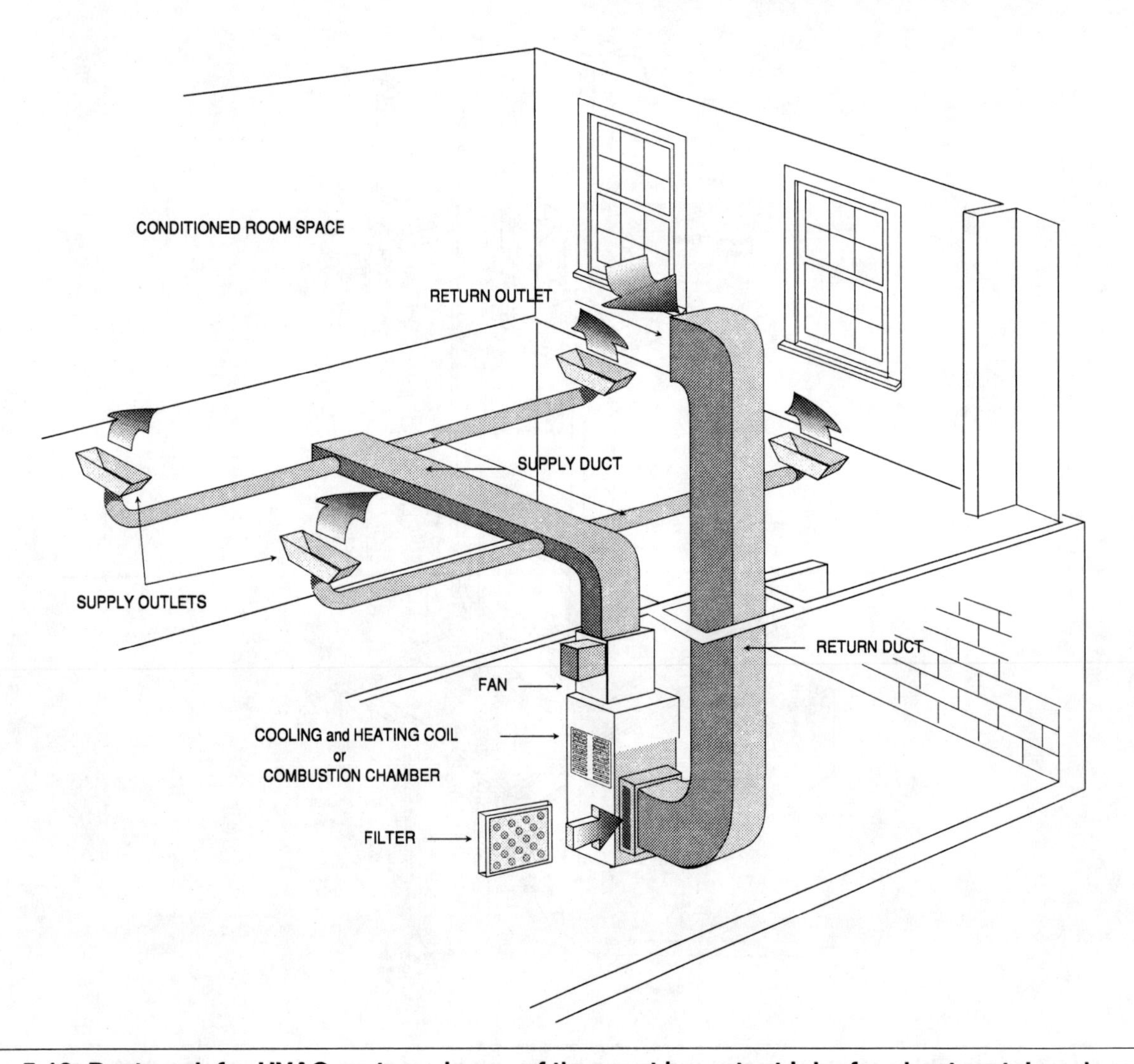

Figure 5-10: Duct work for HVAC systems is one of the most important jobs for sheet metal workers.

The wrong size or type of material for a given project will hurt the design. These trade workers must also be skilled at designing and making geometric shapes, so they must have a good knowledge of math, including plane and solid geometry.

Using blueprints and other construction plans, sheet metal workers determine the amount and shapes of sheet metal required for a job and, if necessary, cut the metal to fit the specifications. Much of this work is done on site; however, some can be done in a sheet metal shop and then transported and installed. Sheet metal workers also construct aluminum siding, metal roofing, and gutters. They use such tools as riveting machines, drills, hammers, soldering guns, and clamps.

Apprenticeship programs for sheet metal workers are four years long and require a high school diploma. Apprentices receive on-the-job training as well as classroom instruction in blueprint reading, math, and drafting. In coming years, the Bureau of Labor Statistics estimates that the construction industry will employ more than 260,000 sheet metal workers. Many skilled sheet metal workers become design and layout specialists. Others become supervisors of sheet metal crews for large construction companies.

WELDERS

Welding is the process by which objects are united by heat or pressure. Welders in the construction industry are classified as either skilled or unskilled.

Skilled welders are masters at their trade. They understand the welding properties of numerous metals, and they can weld objects or fixtures to specification from drawings. Unskilled welders work on assembly lines for large industrial plants and have a very basic understanding of welding properties.

Practically every trade in the construction industry needs the skill of a welder.

There are three types of welding: arc, gas, and resistance. The skilled welder understands each type and knows what type is best to use on a specific object. For example, in arc welding a welder uses an electrode to join entire objects together. The electrode shoots electric current through the objects, causing them to melt.

Gas welding, on the other hand, uses an oxygen or acetylene torch to ignite a welding rod and weld edges of metal together. This form of welding is most commonly used in construction.

Resistance welding is normally used in large industrial plants that produce numerous objects and materials in a short period of time. As in arc welding, an electrode is used, but all the welding is done by machine.

There are no formal education requirements for welders. Most enter the trade by taking welding courses in high school and continuing as unskilled welders in an industrial plant. Skilled welders often study welding at a technical school. Or they learn their trade during years of on-the-job experience working with other skilled welders. Some industrial plants and contractors require welders to pass a skill test before hiring them.

The welding trade is anticipated to grow by many tens of thousands in the coming years.

Chapter 6
Specialized Trades

Besides the basic construction trades, there are dozens of specialized fields from which to choose a career. This chapter describes some of these specialty areas.

FLOOR FINISHERS

Flooring is one of the last features of construction to be completed in a project and provides a finish to the structure's interior. Ceramic tile is the most common flooring used in all types of construction, as well as the oldest. Records indicate that tiles were used as early as 4,000 B.C. Tiles are produced by combining a variety of raw materials at high temperatures to create a glass-like finish known as glazing.

Tile setters make sure that the surface where the tile will be placed is clean and free from cracks, abrasions, and other deformities, and they must correct any problems before work begins.

Tile setters determine how much tile they will need for an installation, decide what type of adhesive is needed, and cut and fit the tile if it has not been precut. Incorrect tile measurements waste time and money because cutting and fitting tile is a difficult task.

Choosing the correct adhesive is also important and is determined by the type of base—concrete, wood, or plasterboard—and the tiling location. For example, the adhesive used in bathrooms must be able to withstand moisture.

Floor finishers are often expected to select and design tile patterns, and they should, therefore, have some artistic ability.

Oak floors are often used in residential construction. Floor finishers must know how to lay tongue-and-groove flooring, as well as how to sand and refinish these floors.

Apprenticeship programs for floor finishers last about three years or 6,000 hours, and cover all types of flooring—including wood, linoleum, and carpeting—as well as all aspects of installation, from sanding and finishing wooden floors to measuring and laying carpet. Some apprenticeship programs, however, focus mainly on tile installation.

As in many other trades, qualified floor finishers learn the craft from other experienced finishers. Expert floor finishers do receive some formal training during their career.

Practically every building constructed will require the services of the flooring trade. The heavy-duty tile floor shown here was expertly laid and will last for decades with little maintenance.

ROOFERS

Roofers install roofing on residential, commercial, and industrial buildings and often work in crews if the project is large.

The most common types of residential and commercial roofing materials are metal decks, asphalt shingles, slate shingles, and cedar shingles. Industrial structures use precast concrete decks and poured lightweight concrete decks. In all cases, roofers must select the appropriate roofing materials and install them correctly to withstand the effects of temperature, water, and humidity.

Roofing materials are applied to sheathing—a base of plywood or metal sheeting that is laid on top of the roof support beams. The roofer waterproofs this base with tar or asphalt and applies shingles. As with most trades, the roofer must be capable of coordinating and communicating with other trades working on the project.

Roofers do not usually need a license or formal training, but they must be able to communicate and work well with other trades people on a project. There are training programs for roofers that train workers on site and cover waterproofing, materials and tools, job estimating, and material ordering.

Besides roofs, these workers sometimes waterproof basements and wall partitions.

DEMOLITION WORKERS

When a building is no longer fit for use or renovation, it must be destroyed by a qualified demolition crew. These workers use explosives and wrecking machines to do their job. Some demolition workers operate heavy equipment; others handle explosives.

The most difficult part of this trade is determining how a structure should be destroyed. Demolition workers must consider the makeup of the building (i.e., brick, wood, concrete); the environment; the location (i.e., urban or rural); local ordinances regarding explosives; safety precautions for workers, residents, and other structures; and tools and equipment required to get the job done.

If explosives are not practical or permitted, structures must be destroyed using heavy earth-moving equipment such as cranes and balls and chains. The building must be emptied of all materials that might affect the crew's ability to destroy it. When explosives are used, demolition workers must determine the minimum amount needed to destroy the building. Once the building is destroyed, the debris must be removed with the help of earth-moving equipment, picks, shovels, sledge hammers, and wrecking bars.

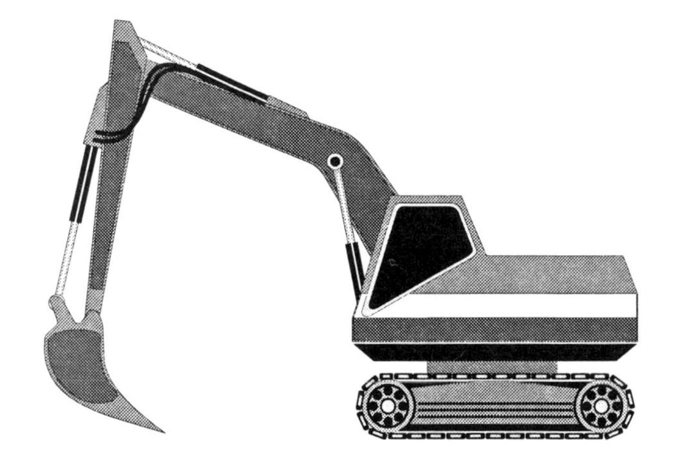

Heavy-equipment operation is one of the job requirements of a demolition team.

Demolition workers normally enter the trade as helpers or laborers, assisting trained demolition experts. Helpers transport explosives to demolition sites, prepare holes to set the explosives in, and connect wires and fuses for setting the explosives off. It takes years of experience to learn how to properly destroy a building with explosives—what type of charge to set and how much explosive to use. Blasters or explosive experts must be licensed. In order to get a license, many states require these workers to pass a written exam and to have letters of reference from licensed explosive experts.

Experienced demolition workers can earn high salaries and can advance to supervisory positions.

Elevator Construction Process

Installation of Guide Rails

Setting Hoisting Machines, Frames and Platforms

Attaching Cable Car to the Base Shaft

Installation of Body and Roof of Cable Car

Electrical Wiring of the System

Testing the System

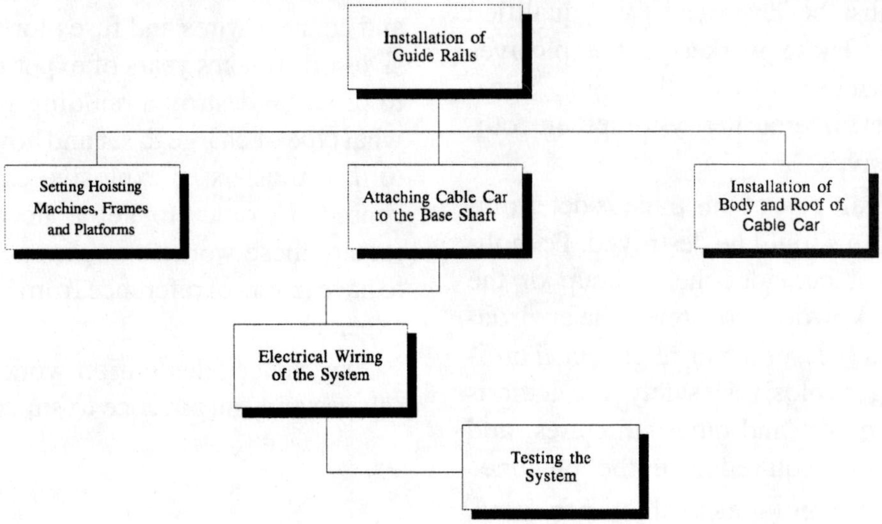

Figure 6-1: Elevator construction process.

ELEVATOR CONSTRUCTORS

Elevator constructors construct and install elevators, dumbwaiters, and escalators. (Figure 6-1 explains the general process for installing elevators.) Elevator repair workers inspect, maintain, and repair elevators, as well as modify older elevator systems to keep up with changing technology, safety codes, and regulations.

Elevator constructors and repair workers are typically trained in technical schools. Applicants are expected to have a high school diploma and a good understanding of math and physics. These programs are usually four years long and combine on-the-job training with classroom instruction. Students can work as helpers after six months in the program.

Most jobs for elevator workers are in large cities, such as New York and Chicago, where skyscrapers are common. Elevator constructors and repair workers almost always work as a part of a crew. They must be able to lift heavy objects and work in tight places, such as elevator shafts. Experienced elevator workers can become foremen of a crew or supervisors for large elevator manufacturing companies. They can also become independent contractors.